T0272513

The Great River

The Great River

The Making and Unmaking of the Mississippi

BOYCE UPHOLT

W. W. NORTON & COMPANY
Independent Publishers Since 1923

Copyright © 2024 by Boyce Upholt

Lucille Clifton, excerpt from "the mississippi river empties into the gulf" from *How to Carry Water: Selected Poems*. Copyright © 1996 by Lucille Clifton. Reprinted with the permission of The Permissions Company, LLC on behalf of BOA Editions, Ltd., boaeditions.org.

All rights reserved
Printed in the United States of America
First Edition

For information about permission to reproduce selections from this book, write to Permissions, W. W. Norton & Company, Inc., 500 Fifth Avenue, New York, NY 10110

For information about special discounts for bulk purchases, please contact W. W. Norton Special Sales at specialsales@wwnorton.com or 800-233-4830

Manufacturing by Lakeside Book Company
Book design by Chris Welch
Production manager: Julia Druskin

ISBN 978-0-393-86787-9

W. W. Norton & Company, Inc., 500 Fifth Avenue, New York, N.Y. 10110
www.wwnorton.com

W. W. Norton & Company Ltd., 15 Carlisle Street, London W1D 3BS

1 2 3 4 5 6 7 8 9 0

For my father,

my first guide to the wilderness

it is the great circulation
of the earth's body, like the blood
of the gods, this river in which the past
is always flowing. every water
is the same water coming round.
everyday someone is standing on the edge
of this river, staring into time,
whispering mistakenly:
only here. only now.

—*Lucille Clifton*, "the mississippi river
empties into the gulf"

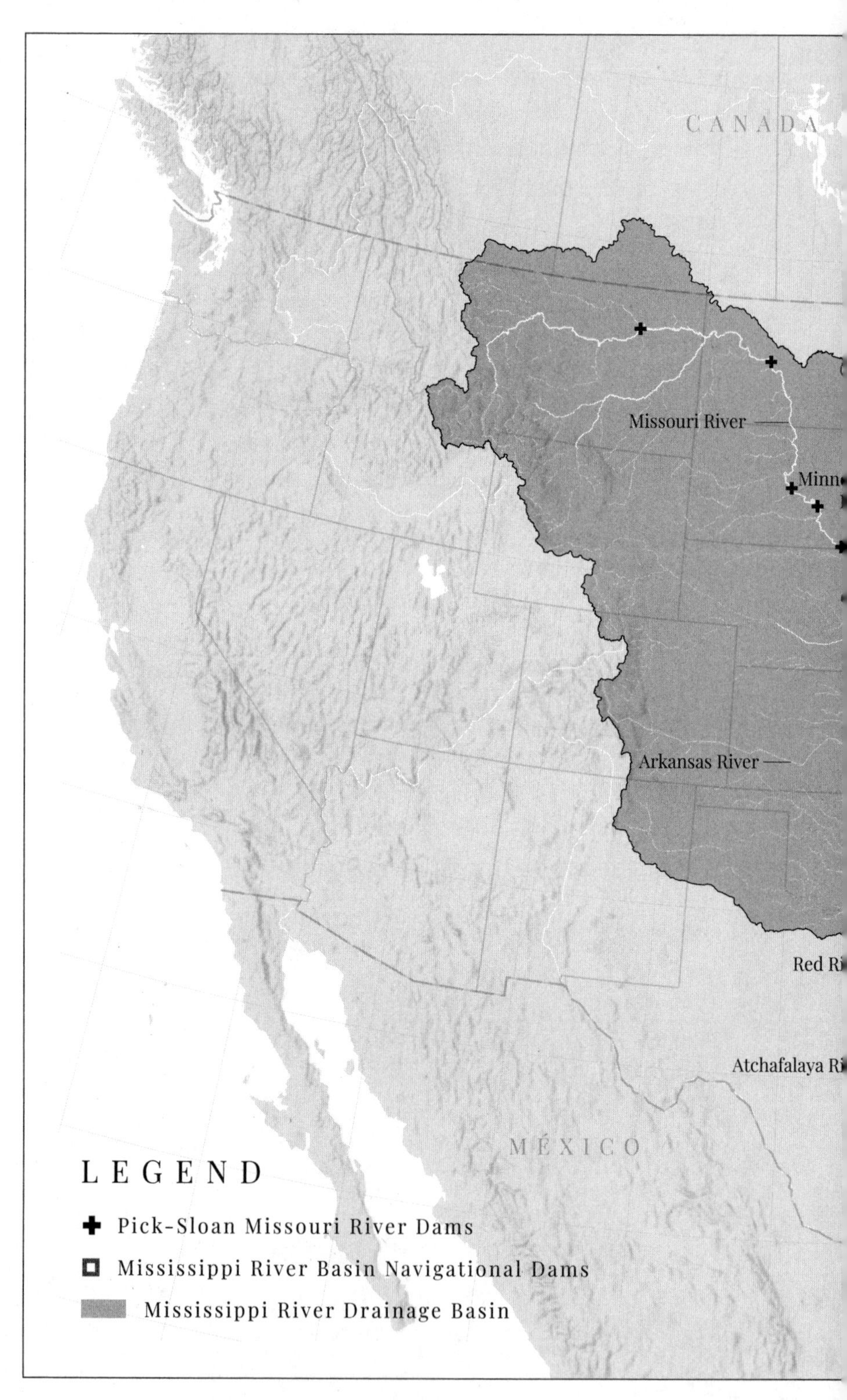

Key geological features of the Mississippi River watershed.

Minneapolis
Wisconsin River
Chicago
Illinois River
a River
Wabash River
Pittsburgh
Cincinnati
Ohio River
St Louis
Louisville
Cumberland River
White River
Nashville
Tennessee River
Memphis
Greenville
Yazoo River
Vicksburg
New Orleans
0
125
250
500 Miles

Contents

Part III: THE UNMADE MISSISSIPPI

List of Maps

Night swimming in the Old White River backchannel (Desha County, Arkansas).

Introduction

A Most Magnificent Spectacle

The tangle of water that we know as the Mississippi River has borne many names through the years. The Dakota called it Ȟaȟáwakpá, "the River of the Falls," because in the heart of their territory the water plunged over a ledge of limestone, dropping twenty feet, coiling into spires of mist. Several names once used across the South live on now in the notes of early European explorers, who did their best to transcribe what they heard: Chucagua, Tamaliseu, Tapatu, Malbanchaya. These men also conjured their own name: El Rio del Espiritu Santo, "the River of the Holy Spirit." The French came next, wandering into the watershed from the Great Lakes, and tried out several options—the Saint Louis River, the River Colbert—before they found one that stuck. *Misi-ziibi* was a word they borrowed from the Ojibwe people, whom they had met in the north country. The French wound up being the first Europeans to traverse the river's full length, and as they traveled downstream, Catholic priests and backwoods fur traders carried this northern name with them, forging a lasting idea.

There are varied ways to translate *Misi-ziibi*: "Big River," say, or "Long River." I prefer "Great River," if only because it reflects the facts on the ground. The Mississippi River drains more than a million square miles, an expanse that encompasses 40 percent of the continental United States: all of seven states, parts of another twenty-five, and a small scratch of two Canadian provinces. More than a hundred thousand named waterways pour into this river, plus countless creeks too small to make the maps. Then there are our artificial channels, city storm drains and sewer lines, canals and ditches on the edges of farms—half a continent of culverts and pipelines carefully constructed to send water on downstream so it becomes someone else's problem. All told, the Mississippi River spits 140 cubic miles of water into the Gulf of Mexico each year, enough to cover the state of Louisiana to a depth of nearly fourteen feet.

The line of water that runs south from Minnesota and bears the label "Mississippi" is just one of several significant contributors. Other tributaries can claim grander superlatives. The Missouri traces a route nearly twice as long into St. Louis, the point where the two rivers meet. The Ohio River, which joins the party another two hundred miles downstream, is shorter but mightier: where its waters enter, at the bottom tip of Illinois, the combined rivers double in flow. Here begins the river I know best. The Lower Mississippi, as geologists call it, is distinct from its tributaries in nearly every way, even in psychic heft. I've heard stories of through-paddlers who, after traveling more than a thousand miles from Minnesota, give up when they reach Cairo, the soggy town built atop a spit of land between the Ohio and the Mississippi. They're overwhelmed by the size of these conjoined waters. Even when the river sticks within its banks, the channel here stretches a mile across.

When you sit upon the Lower Mississippi's surface in a canoe, you realize this river has its own topography—crests and valleys, feet high; curls and curves and whorls; wide, still boils. The long light at sunset and sunrise reveals a labyrinth of ripples. On some mornings, the water lifts into mist so thick you realize there is no end to the air and no beginning to the water, so your boat floats upon and within the river at once. Nor does this river have any real bottom. When the engineer and inventor James Buchanan Eads sank himself into the Mississippi near Cairo to salvage lost cargo, his feet never found solid purchase. There was just a rushing current of increasingly dense mud. Eads's rival, Andrew Humphreys, conducted the U.S. Army's first comprehensive study of the river and noted eddies that spanned half its width and surged upstream at seven miles an hour. Humphreys declared the river a "turbid and boiling torrent, immense in volume and force."

A few sentences later he added, rather hopefully, that all this water was "governed by laws."

The first written records of the Mississippi River come from the men who accompanied Hernando de Soto, a Spanish conquistador, in his sixteenth-century ramble across the continent. They were searching for gold, which they never found. Instead, their journey turned into a death march: the Europeans delivered death to the Indigenous towns they found, and the Europeans suffered death, too. Soto himself succumbed to a fever, and his soldiers dropped his body into the river in 1542.

The next winter the expedition's surviving members holed up in a village they'd previously ransacked. For months, they watched the river's water rise, expanding across the floodplain, until the pool spanned more than sixty miles. "It was a most magnificent spectacle to behold," a historian wrote a few decades later, based on his interviews with the soldiers. For as far as they could see, "nothing was visible except the pine needles and branches of the highest trees."

Back then more than fifteen thousand square miles went underwater in big flood years, when the upstream rivers flushed the snowmelt and spring rains downstream: Such is the expanse of the Mississippi's southern floodplain, the low-lying land that is subject to the river's caprice. The story of a river is always also the story of its floodplain—or at least it should be. That's because the two are connected, or they should be. Floods once linked the river to backwater lakes, providing fish with quiet corners to lay their eggs. Floods delivered layers of rich soil that made it easy for the river's first inhabitants to harvest marshelder and chenopod and knotweed and maygrass—ancient crops that now grow only in feral form—and later squash and beans and corn. When Soto arrived, wetlands covered at least 24 million acres along the southern river. Combined with the adjacent gulf marshes, this was among the world's largest expanses of swamp, critical to the biodiversity of the entire continent.

These swamps were critical to the continent's human cultures, too. Soto's men rambled from Florida to the Carolinas to Texas; nowhere did they find so much wealth and power as along the lower river. The chief of each village lived inside a home atop a carefully engineered mound—a form of spiritual architecture that was born in the river's swamps five thousand years earlier. The first of these earthworks appeared in Louisiana, where people gathered together to build complexes of rings and ridges and mounds. The size of the monuments grew through the millennia, and by 1100—five hundred years before Soto arrived—the river swamps in Illinois featured a sprawling city with 120 pyramids, the tallest rising a hundred feet above the surrounding creeks and bayous and sloughs. Long before any European king could even dream of this watershed, it already had its first superpower.

We tend to think of the embryonic United States of America as consisting of thirteen coastal colonies. But the country—the river's modern superpower—was dragged into being in large part by men who scrambled over the Appalachian peaks and followed the creeks they found on the far side. The British king had forbidden western settlement, but these men were drawn to the Mississippi, determined to seize land and therefore wealth—so determined that they went to war, launching a revolution that birthed a new nation.

After the Revolutionary War, the Mississippi watershed became white America's first frontier, a Wild West before anyone dreamed of deserts or Rocky Mountains. "The men of the Mississippi Valley compelled the men of the East to think in American terms instead of European," the historian Frederick Jackson Turner wrote in his seminal book *The Frontier in American History*. "They dragged a reluctant nation on in a new course." But while the expanse of the watershed, and the rich soils of the floodplain especially, lured land-hungry settlers, the river itself inspired more caution. William Alexander Percy, an early twentieth-century poet who made his living overseeing his family's massive plantation, called it "the shifting unappeasable god of the country." The Mississippi was the undeniable life force of the cotton kingdom, the builder and underwriter that had delivered the soils that made Percy's riches possible. It was also the menacing force that could wipe his property away.

For centuries, engineers worked to tame this god. The first levee along the river was completed in 1720 in New Orleans (or the Île d'Orléans, as its founders sometimes called it, the Island of Orleans, because it was surrounded by so much water). It was a simple technology, a waist-high wall built from boards and tree trunks, buttressed from the back by a mound of soil. As the levee grew beyond the city, this fortification turned even simpler: no wood at all, just a pile of dirt. A century later, once the Americans claimed the river, the project of conquest grew more technological—and more violent. Crews packed dynamite into the river's rocky reefs, blowing open navigable channels. They dispatched steam-powered battleships to smash through sunken trees. Later, hydrologists assembled concrete models, including one as big as a small village, so as to comprehend the river's misbehaviors. Through it all, various camps of engineers squabbled about science and theory and who should be in charge.

The winner of that fight was a bureaucratic behemoth suited to the might of this watershed. The U.S. Army Corps of Engineers once dug a tunnel in Alaska to study the effects of excavating through permafrost. The agency built a city in Saudi Arabia to house that country's troops. But the Mississippi watershed is its masterwork. The Corps has built dozens of locks along the Upper Mississippi and the Ohio, rendering both into watery staircases so that towboats can cart the nation's grain and coal. The Missouri features six massive dams that together create the largest water-storage system in the country.

But it's on the Lower Mississippi, the unruliest of the nation's rivers, where the Corps's ambitions reached their peak. The Mississippi River and Tributaries Project, as the system of infrastructure is officially known, is "arguably the most successful civil works project ever initiated by Congress," according to the Corps. There are no dams on the southern river—there is too much water for dams to be necessary—but hundreds of rock dikes direct the flow of the water, and thousands of miles of concrete padding along the banks prevent erosion. Fourteen engineered shortcuts are meant to speed the water along and thereby decrease floods. Four different "floodways" offer release valves for when the water rises too high.

The crowning feature of the MR&T Project is the modern levee, which now snakes along the southernmost thousand miles of the Mississippi, opening in only a small handful of places to allow smaller rivers to offer their tribute. The world's longest levee stretches along the river's west bank, from the mouth of the Arkansas River to the headwaters of the Atchafalaya, an unbroken wall of earth that spans 380 miles. The only longer human-made landform on the planet is China's Great Wall. Levees have yielded a kind of landscape new upon the Earth: a dried-out floodplain. It's possible now to walk just a stone's throw from the continent's largest river and have no idea you're near water at all. You stand instead in a neat and tranquil neighborhood, or amid a dusty field of soybeans, protected by the quiet green slope of the wall, which stands forty feet tall or more.

I grew up far from this river, amid a sprawl of suburbs in Connecticut, a place where the word *Mississippi* was just a four-syllable approximation of the length of one second, convenient for counting time in a game of tag. Even after footloose happenstance delivered me into the walled territory along the river—I took a job in rural Mississippi working in struggling schools and began to eke out a life as a writer on the side—I paid little attention to the nation's big river. Its waters were irrelevant because they were invisible, hidden behind the wall.

Then I took on a magazine assignment, to write a profile of a man named John Ruskey. Like me, Ruskey had wandered into Mississippi from far afield, Colorado in his case, lured by the chance to learn blues guitar at the feet of the masters. Eventually, Ruskey took an office job in a small-town museum to pay the bills. A child of the mountains, he found the hours trapped behind a desk were rather claustrophobic; for relief, he went out to the river, exploring its islands and backchannels by canoe. By the mid-1990s, he was well known as a riverman, so when a Belgian tourist asked how to see the Mississippi, someone suggested he hire Ruskey for a tour. Two years later Ruskey founded the Quapaw Canoe Company, the first guiding service anywhere on the lower river.

By the time I met him in 2015, Ruskey had been working for years on a mile-by-mile guide called *Rivergator*. Intrigued by the project, I talked my way onto the crew for Ruskey's final reconnaissance expedition. We were a group of eight, of whom I had the least experience on a big river: none at all.

The morning we launched, the Mississippi was so high that it had swallowed the road to the boat ramp. But someone had dug a ditch alongside a farm road, which, thanks to the flood, had become a tiny tendril of the country's great river system—a canal that led out onto the flat sheet of water spread atop the field. After a mile, this pool merged into a sluggish little river known as the Buffalo, which after another two miles joined the mighty Mississippi. The surface of the ditch was scummy and green, laden with whatever chemicals had run off the surrounding farm fields and cattle pastures. A snake shimmied across its top. Ruskey grinned. Then he jumped in: his ritual river baptism.

On the big river, the flood left no landscape, just the tips of barren willow branches reaching up from brown water. Trash studded the few visible lines of riverbank: washed-up buoys, bottles and cans, food wrappers; once even an overturned refrigerator. A few times an hour, a towboat gnashed through the pool, carrying its loads of commerce: thirty barges or more packed with grain or chemicals or scrap metal, assembled together in a long, thin line—as much as five acres in total, as big as a small floating farm. After each boat passed, the water remembered its presence, its surface convulsing with waves. Mark Twain once called this a "river of desolation," and to me, on that afternoon, his description seemed apt.

But as we paddled toward the island that would serve as our first campsite, a warm breeze greeted us. We arrived on a strip of sand that rose until it met a wall of willow trees. Our party spread across the dunes. I pitched my tent at the edge of the forest, from which tracks—coyotes, beaver, deer—emerged.

To the east, I could see the glow of lights from the Louisiana State Penitentiary, the country's largest maximum-security prison. To the west, I could hear pickup trucks roaring on a county highway across the water, beyond the levee, in the version of the floodplain that I knew. It was green, yes, but all straight lines and hard edges—an entirely human domain. On that island, meanwhile, sheltered under the vast sky and its stars, hemmed in by the water and the woods, listening to the lapping river, drifting in and out of sleep, I felt like I was in a separate world. This place was windswept and water-blasted, composed of curling lines. It was not a typical wilderness, perhaps, but it was as wild a place as I had ever been.

The French had a special name for the land between the levees, a word that's still used today in the South: *batture,* which comes from *battre*—"to hit, to beat, to batter."* It's a name that makes sense, because these are flood-battered lands. The batture is a place that tends to defeat any human aspirations. Build a cabin there, or plant a farm, then wait a decade. All evidence of your presence will be washed away. Indeed, because the Mississippi is always shifting, the land you once stood upon might be gone, swallowed by the river.

That first trip on the river sparked an obsession, first with the batture, then with the whole of the watershed. In the decade since, I've assembled a miniature library, of books on the histories of the river's great cities, on its Indigenous customs and lifeways, on the travails of the first European explorers to encounter these waters. I've got memoirs of young hippies who ran away from society to live in the swamps and scholarly studies of the squatters who occupied the "shantyboats" tied to its banks. I've logged more hours in archives than I care to count, thumbing through the crumbling pages of out-of-print travelogues. Twain once said that the

* The word's French roots are somewhat obscured by its pronunciation, which, like many imported words, has been adapted to the Southern accent. *BATCH-er,* you'll usually hear.

Mississippi is well worth reading about, and he is right, but my time in those carrels always reminds me it's well worth seeing, too—which is why I've spent hundreds of days, thousands of hours, paddling the river, sleeping on its banks through snowfall and blazing heat and thunderstorms strong enough to blow down trees.

I've found a lot of beauty out there. I've also found reasons to worry. When I skimmed across a frigid lake in a johnboat with a fisherman—the only river fisherman I could find still working in the state of Mississippi—he hauled up nets full of undersize buffalofish. They were being starved out, he told me, by invasive Asian carp. When I toured the marshes along the river's coastal edge—which supply a billion pounds or more of seafood in some years, nearly a third of the landings in the continental United States—my guide pointed out the pilings at the sites of former fishing camps. They'd been abandoned and demolished because the marsh itself was disappearing. Every day, on average, eight hundred thousand square feet of these wetlands dissolve into the ocean. Pipelines and dredgeboats carry mud to the edges of the Louisiana coastline, constantly rebuilding the buffer of wetlands that protect inland towns and cities against storms. On one early trip, I traversed an old oxbow lake with a biologist who was sampling its water for microscopic life. He told me that because our modern engineering has locked the river in place, new backwaters are no longer forming. The old backwaters, meanwhile, are filling, as they naturally do, thanks to the endless tide of mud. No one knows what will happen as these slow backchannels, the spawning grounds for so many species, disappear.

The dry side of the levee has not fared much better. Drained and cleared, the river's swamps have become an incredibly productive swath of farmland, but also a landscape denuded. Most of the local species have been extirpated, often intentionally. Those that we want—soybeans and corn and cotton and rice—are kept alive only thanks to pumps and chemicals and genetic engineering. Nor has this regime produced an economic windfall. Nearly every county along the Lower Mississippi River is marked by stark rates of poverty.

The river, meanwhile, remembers its old ways and keeps on reaching outward, flooding its former domain. In an eight-year period early this millennium, the river hit its "ten-year flood stage" seven times in Mark Twain's hometown of Hannibal, Missouri. Run a million simulations, and such a sequence should only happen once. But this is not random. The river we've built is coming apart.

Back in the 1940s, as the Corps of Engineers was entering its greatest bout of river-building mania, the Argentine writer Jorge Luis Borges published an enduring fable: the story of an empire whose cartographers grew too detailed. Their map of a province would be big enough to cover an entire city; the map of the country unfurled across a province. Eventually, these cartographers produced a map so large that it covered the whole of the empire it was supposed to depict. Future generations, for obvious reasons, decided that this document was worthless. So they let it rot away, until just a few tattered bits remained in the desert, where "Animals and Beggars" lived atop the scraps.

As I've studied the Mississippi, I've come to think of the modern river as something like this map: it's an imagined canvas that we've stretched atop the geological frame of the continent. The names we use are temporary, labels we've adopted just for this moment. This is not to say the locks and levees aren't real, or that they have no consequences. It's more that, like a map, they aim to freeze an ever-changing river in place; like a map, the concrete and steel will wither with time. And beneath the infrastructure, another Mississippi River persists.

The Great River is a tale of both rivers: of the map and of the territory. Of a river that has been the victim to some of the world's most intense and elaborate engineering, and of a river that is as wild a place as you can find on Earth. I have aimed in this book to be as rigorous as any scholar, but I am not a historian, and this book is not a standard history. To peel through the layers of the river, I found that at times I've had to abandon the comforts of chronology—though this, I think, makes for a story truer to the river itself. The Mississippi never sticks to its banks. It is always winding, finding new routes. I hope that as I have wound, I have carved out paths that both surprise and reveal.

I hope, too, that you will see the beauty that persists despite our mistakes. And I hope most of all that you will be tempted to come and join me on the rivers across this great and sprawling watershed. Your presence—all of ours—is what the Mississippi most needs.

Part I

ALLUVIAL CHRONICLES

TIME IMMEMORIAL TO 1803

The river's headwaters at Omushkos (Itasca County, Minnesota).

1

Searching for a River

White explorers confront a watery torrent

In 1803, when Thomas Jefferson purchased the territory of Louisiana, what he really acquired was a watershed: nearly all the land that drained into the Mississippi River, or at least all the land that the United States had not already claimed. But what was out there? Jefferson's library included more than a dozen firsthand accounts of the river basin—the journals of French explorers, the notes of British surveyors. Yet even he had little clarity. The best maps of the era showed mostly empty space. It was all just "wild land," as one newspaper editor complained.

So Jefferson organized a series of expeditions. The most famous launched from a riverbank just north of St. Louis in 1804. As Meriwether Lewis and William Clark poled up the Missouri River with a fleet of keelboats, they became the first white settlers to document the grizzly bear and the black-tailed prairie dog and the pronghorn. They smoked tobacco with Lakota men amid the vast, unfurling prairies. They paddled into a dark canyon in the Rocky Mountains and eventually left the Mississippi's watershed behind.

Zebulon Pike trekked north in 1805, onto the Upper Mississippi, where he was forced to drag sleds atop an iced-over channel. After he was assisted by the British fur traders who had already come to know the land, he shot down the Union Jack flag at one of their forts. This rude act was ceremonial: one of his missions was to assert his nation's ownership of this territory. It's safe to assume that as soon as Pike left, the old flag was restored. His second mission, to find the Mississippi's headwaters, wound up a failure, too. Pike turned around seventy-five miles short of the lake that today is considered the river's official source.

Soon thereafter Pike received orders to mount a second expedition, an overland voyage to find the headwaters of both the Red and the Arkansas rivers. This was the most mysterious land in the Mississippi watershed, inhabited, according to some stories, by unicorns and giant serpents. Eventually, in Colorado, Pike trudged through waist-deep snow up what he thought was the tallest peak in the Front Range. But he did not reach the summit, and as it turned out, he'd picked the wrong mountain; a later explorer would find a higher peak. Afterward Pike steered south, where he ran afoul of Spanish soldiers. They detained him in Spanish territory for months.

These forays up the Mississippi's tributaries continued for decades, a careful cataloging of a vast network, a watery skeleton that runs through the nation's heartland, linking many different terrains. Where the rivers drained mountain slopes, the water ran cold and fast and crystalline. Down the Rocky Mountains, the streams passed spruce and pine trees; through the Ohio Valley, river birch and walnut. Across the plains of Nebraska and Kansas and Oklahoma, the rivers—dark lines of mud—wound through floodplain forests of cottonwoods and willows. In 1820, on a journey on the Upper Mississippi, the geologist Henry Rowe Schoolcraft noted prairie grass and willow trees, as well as the "humbler growth" of sumac and sarsaparilla.

Schoolcraft had joined an expedition looking to finish what Zebulon Pike had started: to find the ultimate source of all this water. His party went farther upstream than Pike's, reaching a glacial lake fed by creeks that were all too shallow for their boats. Still, Schoolcraft was not satisfied. The local tribes said that when the waters rose, a canoe could push farther. So twelve years later Schoolcraft hired an Ojibwe guide to lead him up another creek. The guide, Ozaawindib, delivered Schoolcraft to the little pool that the Ojibwe know as Omushkos, or Elk Lake. Schoolcraft made a new name. He slammed together the Latin words for truth (*veritas*) and head (*caput*), then lobbed off the outside syllables: Itasca, he called the place, implying that this was the great river's true head. The lake sits amid the sandy-soiled hills and prairies of Minnesota, surrounded by pine forests. Schoolcraft described it as "one of the most tranquil and pure sheets of water of which it is possible to conceive."

Schoolcraft was right that this is a beautiful place. Still, for a long time I was unwilling to visit. His name, Itasca, "true head," was what kept me away. It's a double lie: first because of its pseudo-Indigenous sound, which pretends to be ancient, as if the word were passed down through generations; and then because of its literal meaning. There is no true head to this system.

The peculiar obsession with pinning down the river's beginning persisted for decades after Schoolcraft's journey. In 1884, another man asserted that Itasca, too, was just one more stopover. He'd found a smaller lake that should be given the honor, he said. Amid the debate that followed, the Minnesota Historical Society pushed for a law that would enshrine Schoolcraft's choice as the official headwaters. The law was intended, as the historical society put it, to ensure that the river's "earliest explorers be not robbed of their just laurels." The controversy prompted one more survey of the headwater region, this one conducted without Ojibwe guides or any guide at all—as if to mark the leap from hearsay and folktale into true science.

The resulting document is rather muddled. Its author, Jacob Brower, conceded that the Missouri River begins at a higher elevation and a farther distance from the sea; in order to justify Minnesota as the true source of the nation's mighty river, he claimed (with little evidence) that "from the earliest times coming within human knowledge, prehistoric, aboriginal, Spanish, French, English and American," everyone had traced the river north instead of west. When it came to the local headwaters, Brower conceded that Itasca, née Omushkos, was not the true source of the river. But none of the smaller lakes could be singled out as the source, either, since they were multiple. Because it provided a convenient, central location for visitors, in 1891 the Minnesota legislature made Itasca the center of its first state park. And that is how the world came to decide where North America's great river began.

Once so enshrined, the place was no longer considered sufficiently beautiful. Its egress from the lake was a muddy outflow, "a sight that is not becoming to such a great river," as one state official put it in 1933. So a work crew was dispatched to lay a hidden concrete dam beneath the mouth; on top, they set a line of boulders. The nation's great river now featured a suitably majestic beginning: a small set of rapids, where the water sparkles as it runs in thin sheets atop the rocks. The subsequent channel is a human construction, too, carefully designed to maintain a consistent width. Because the river's final outlet had already been reworked—jetties were installed in the 1870s, clearing out shoals for the sake of the passing vessels—the river had become, quite literally, engineered from its head to its mouth.

The location of the river's beginning is not the only point of confusion; it's just as hard to determine when the river was born. One place to start the Mississippi's story is 600 million years ago—long before the birth of North America—when a supercontinent began to split. This violence left behind a network of subterranean scars, including one known today as the Reelfoot Rift. As North America split free and drifted toward its current home, the rift was exposed to heat escaping upward from the earth's core. The crust cracked and downwarped and ocean water filled the resulting trough. Thus 70 million years ago, the Gulf of Mexico curled upward into a horn-shaped "embayment" four hundred miles tall. Before there was a Lower Mississippi, then, its valley was an arm of the ocean. There are still places where you can find shark teeth buried in the soil. The Chitimacha, a tribe in Louisiana, seemed to intuit this history: they tell a story about the crawfish who, following the instructions of the Great Creator, dove down to retrieve some mud, and so built the world. Geologists have a similar tale, but they credit the nascent Mississippi with delivering the materials.

As the water flowed across the ancient continent, it wore away the dirt, which it carried downstream and dropped into the ocean. The coastline extended slowly southward. After 30 million years, enough dirt had been dumped into the Mississippi Embayment to form what we know today as central Louisiana. To our modern eyes, this landscape would have been wildly unfamiliar. The terrain stood two hundred feet higher than today's floodplain, and the ancient river, which drained the land as far north as central Saskatchewan, carried six to eight times more water than the modern Mississippi.

Then came the ice. Two million years ago, glaciers crept south from the poles, then melted, retreating north again—then descended and retracted again, four cycles in total, which, speaking geologically, ended only just yesterday. The ice never reached beyond St. Louis, but its ebb and flow shaped the lower river. By trapping water, the glaciers lowered sea level; since water seeks equilibrium, these low seas spurred the ancient Mississippi to carve a deep trench. (Now just one line of the ancient landscape, known as Crowley's Ridge, rises amid the otherwise lower floodplain.) In the eras when the ice was melting, the flow turned into a braided river. The Mississippi featured a tangle of channels, weaving between islands and sandbars, running wide and shallow and fast. Huge torrents of water poured off the glaciers; hunks of sand and gravel tumbled along the riverbottom and dropped out in trailing piles.

The glaciers also shaped the tributaries. The ice formed a dam across the northern continent, creating a zone where no river could flow, forcing the Ohio and the Missouri into rough approximations of their current paths. And when the ice made its final withdrawal twelve thousand years ago, its backward retreat carved hills and glacial potholes across the Upper Mississippi Basin. The new rush of meltwater, meanwhile, blasted open floodplains. Along the Missouri River, the bluffs sometimes stand eighteen miles apart. Schoolcraft estimated that the Upper Mississippi ran through a valley that spanned two miles. Downstream from St. Louis, after these two rivers join together, the floodplain—six miles

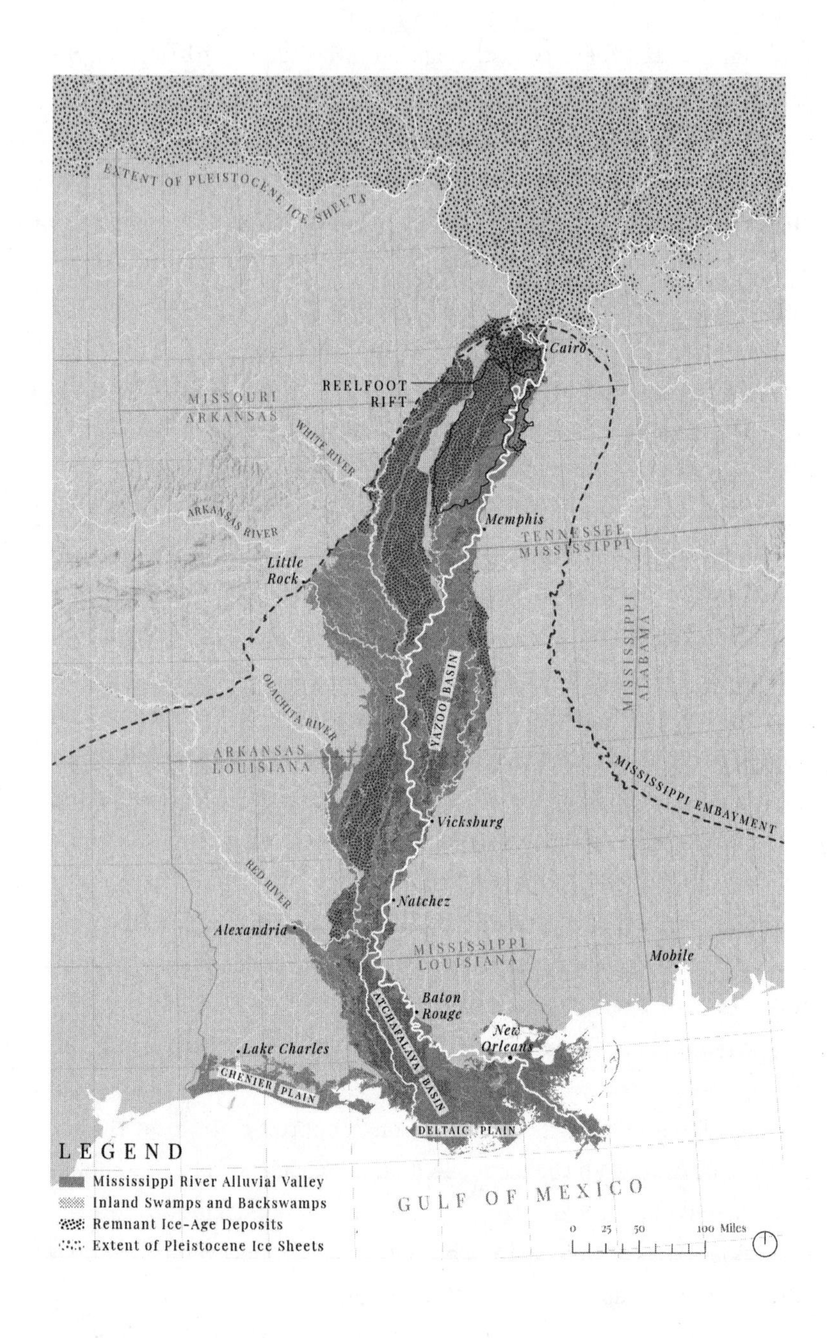

Key geological features of the Lower Mississippi River watershed.

wide, on average—has its own name: the American Bottom. (Charles Dickens visited in the nineteenth century and described it as "one unbroken slough of black mud and water.") Two hundred miles downstream, when the river begins to run into the old embayment—at the head of the Lower Mississippi—the floodplain suddenly expands, reaching ninety miles across.

The Lower Mississippi today, for all its might, is not as violent as the old braided river. The glaciers mostly finished retreating eight thousand years ago, so the rush of water subsided. The river settled into a single channel, a lazier curl of water and mud. This river flows through some of the flattest and lowest land in the world; just south of Vicksburg—four hundred miles above the mouth—the riverbed goes below sea level for good. A river cannot rush through such terrain. It must wind.

The water runs fastest on the outer edge of these bends, since it has farther to travel to keep up. This churn of water tears at the banks until they crumble. Thus the bend grows wider. On the inside of the bend, meanwhile, the slower water drops some of its mud. These forces together encourage the river to make an ever-wider curl. Mark Twain, in *Life on the Mississippi,* devoted a whole chapter to the resulting horseshoe-shaped bends. In some places, he wrote, you could disembark from a boat, and after walking "across the neck, half or three quarters of a mile, you could sit down and rest a couple of hours while your steamer was coming around the long elbow, at a speed of ten miles an hour, to take you aboard again." Sometimes the water itself would take the same kind of shortcut, carving a new channel and abandoning the curve of the horseshoe. (The abandoned miles of river would then become what's known as an oxbow lake.) Somewhere else a new horseshoe would be growing. For thousands of years, the river writhed this way across its floodplain—meandering, in the geological term—inscribing a map of its travels upon the landscape. The "land's slow alluvial chronicle," William Faulkner called the result.

Historical records show the river carved itself sixteen such cutoffs between 1765 and 1884. A keen eye can read the chronicles into the deeper past. The most conspicuous marks of the river's travels are its natural levees. These were built as the river rolled over its banks; upon slowing, the water dropped out its largest bits of mud and sand, building a low ridge—standing little more than fifteen feet above the surrounding terrain—that clasps the river and stretches as much as three miles across. This zone constitutes what's known as the meander belt, marking the breadth of the river's wandering. Occasionally, the river blasts through its natural levee, creating not just a new bend but a whole new meander belt—an alternate route through the valley, dozens of miles away from its old path.

Thus, through the post-ice years, the river carved through much of the older sand and gravel and laid down its own muddy terrain. This sediment—alluvium, as it's known—is literally a distillation of the United States, the rocks and dirt of Pennsylvania and Georgia and New Mexico swirled together by the water and deposited in layers. This was a transfer, then, as the rest of the continent was carried south to build a new domain. Today it's known as the alluvial valley. Unlike most valleys, here the riverbank is the high ground. As you walk away from the Mississippi, the natural levees slope gently down into the backswamp, where slowly unfurling floodwaters have delivered only small bits of silt and clay—though even here, after thousands of years of flooding, the accumulated layer of muck reaches as much as a hundred feet deep.

This dirt is rich. A surveyor traveling in the alluvial valley in 1880, before the land was cleared for farms, cataloged the many trees he found. On the high ridges stood white oak and willow oak; shell-bark and mocker-nut hickories; black walnut and yellow poplar and sassafras; mulberry; Spanish oak; and gum trees, sweet and black, amid smaller undergrowth of dogwood and crabapple and wild grape and buckthorn. As the land dipped into the backswamp, the surveyor found more oaks—cow oak and water oak and red oak—alongside bitternuts and elms, hornbeams, white ash, box elder and red maple, honey locust, holly, and pawpaw, this last an overlooked treasure that lines our riverbottoms and each autumn yields a sweet, custardy fruit.

That list leaves out the king of the bottomlands, the bald cypress. The world's largest bald cypress is also the largest tree of any kind in the United States east of the Sierras. Ninety-six feet tall, it stands a mile from the Mississippi River, in Louisiana, outside the town of St. Francisville. Its trunk stretches seventeen feet across—forty-six feet in circumference, so fat that a chain of eight full-grown men stretching their arms wide would not quite reach around. A bald cypress can tolerate water, surviving years of floods to live for two millennia or more. But most of the ancient cypress were chopped down in a fifty-year spurt, beginning in the late eighteenth century. This wood was too perfect to let stand: it was rot-resistant and pest-resistant, and a single tree could yield over fifty thousand board feet—enough wood to build a dozen small, three-bedroom homes. That the champion bald cypress survived to be fifteen hundred years old is thanks only to its hollow core. There was no good reason to fell this giant.

Animals would come to visit these lands, too, at least when the mud grew dry: opossums and bobcats, cougars and black bears and red wolves and deer. A French crew that paddled upriver in 1699 noted three bison napping on the stretch of the bank that would soon become New Orleans. What brought all these creatures were the easy pickings at the bottom of the food chain, fruit and acorns, mayflies and crawfish and mussels and algae in the mud—the varied abundance of the swamps. For humans, such abundance wasn't always pleasant. The first French explorers complained that as soon as they moved south of the Ohio River, the mosquitoes became unbearable. Decades later, once the French established a colony, they used these mosquitoes as punishment: recalcitrant soldiers were tied naked to trees. I can confirm the severity of the penalty, as I have been caught on the river after dark and found myself drowning in the insects. For much of the year, camping in a tent can feel like a slumber party inside a sauna. This is not the classic American wilderness—not a tall mountain or deep canyon, the kind of place where human beings stand out, lonely, against the expanse. In the swamp, a human being turns lonely in a different way. You become smaller, just one bit of all the shrieking biomass.

One thing you will not find along the Lower Mississippi is rock. One of the first government-commissioned studies of the lower river defined its northern limit as the point where rock appears on both sides of the channel—making this, officially, a world of mud. The riprap that protects the riverbanks today—piles of rock that stop erosion—has to be carted in from upstream.

The edges of the alluvial valley are lined with bluffs. To the east, they stand two hundred feet tall and are built from silty soils blown off the lowlands by ancient windstorms. To the west, it's harder to discern the edge of the alluvial valley, since it's gashed open by major rivers—the Arkansas, the White, the Red—whose own floodplains merge with the Mississippi's. Geologists divide this muddy valley into seven separate sections of lowlands. The largest, and the only one on the river's east bank, is also the most famous. The Mississippi Delta, as it's often called, is known to geologists as the Yazoo Basin.

For the sake of clarity, in this book I'll use the geologists' name. The entire valley along the Lower Mississippi River is sometimes called the Delta, and to add to the confusion, the true geological delta begins only near Baton Rouge, where the river begins to fork apart, forging various paths toward the sea.*

The delta marks the end of the alluvial valley. This is where, as the ice ages closed, the river reached the edge of the continent and began to pour its mud into open ocean. The land built slowly outward at a rate of two to three square miles per year. Because the coarser dirt was dropped first, a shoal formed at the mouth, splitting the water into several branches, which would split in turn. The only arable ground runs in thin strips along the resulting waterways, and rarely tops ten feet. If you walk directly away from the river, after just a few miles the dry dirt gives way to soggy cypress forests. As you move toward the gulf and the water grows saltier,

* This muddled naming does contain some geological truth. A delta, in addition to the place where a river forks, is the place where the land has been built by the river. And that is how the land within the alluvial valley was built, albeit over millions of years, compared to a few thousand years in the case of the modern delta. I like to think that the name Delta persists for northern portions of the alluvial valley because this landscape—low and flat and muddy—is so distinct from anywhere else in the country that it is obvious to everyone that it belongs to the river.

the forest turns to marsh, nothing but sedge and grass, which gives way in turn to the endless gray water of the sea.

If the upriver landscape seems unsteady, the delta cannot be trusted at all. The weight of the latest mud is always pushing downward, causing the land to sink. Hurricanes steam through, ripping the marsh apart. And here when the river smashes through its natural levee to create a new meander belt, the water points in a whole new direction, toward a new mouth. (Often this new mouth overlaps a lost, ancient patch of delta; the latest dose of mud is laid atop ground that has already sunk beneath the sea.) This is a place where the act of creation has never ended. The mud beneath New Orleans is some of the youngest land on the continent, little more than four thousand years old. By way of comparison, much of the surface rock across the midwestern United States was formed during the Paleozoic era, which ended hundreds of millions of years ago. The difference goes beyond orders of magnitude: if you compress all the time since the Paleozoic into a single day, New Orleans's mud begins accumulating seconds before midnight.

The historical record, meanwhile—the record that begins when Europeans show up with their journals and ledgers—takes up less than a second on that clock. Even then, throughout its first several centuries, it offers nothing but sputtering, flash-frame glimpses of the river. Across the Southeast, Hernando de Soto found a collection of walled cities, each typically independent and self-sufficient, though some chiefs built alliances, submitting to the power of a nearby leader. Archaeologists today describe such small societies, where elites and rulers inherited their power, as chiefdoms. Soto and his men noted that the chiefdom cities along the Lower Mississippi River, with their big size and organized armies, were the most impressive they encountered on their continental tour.

This was not yet the era when explorers were educated gentlemen. Picture Soto's men—seven hundred strong when they first arrived in Florida in 1539—as a warring medieval horde. The soldiers supplied their own equipment and received no salary; instead, they would be rewarded a portion of the plunder, which suggests the kind of economic thinking that motivated this force. They were seeking neither some colony nor real estate, just gold. Soto was an unlikable man—a quick decision-maker, with little patience for whatever he deemed foolishness. As he rambled across the continent, he sought wealth by whatever means, murder and mutilation included. Among his key weapons were the dogs that traveled with his army, trained to seek genitals and then disembowel their victims.

After Soto died in 1542, his corpse was wrapped in shawls and weighted with sand, then dropped into the Mississippi River. The survivors, half or fewer of the original army, had had enough of the territory. First they tried to walk overland to Mexico, where there were already Spanish settlements, but couldn't find enough food to steal from the buffalo-hunting societies that lived in the hinterlands between. So they returned to the river. If they could follow it to the gulf, they knew, they could set sail for Spain's island colonies and be back to their own society. The men spent the winter and spring building boats, yielding the first recorded observation of the river's great capacity to flood. They set off downstream in early July 1543.

On the Spaniards' third morning on the river, six hundred miles upstream of the mouth, amid the dense jungle of the alluvial valley, Quigualtam's navy appeared. For more than a year, Soto's men had been hearing the name of this chief. Now they learned why. His soldiers were masterful. The fleet included several dozen war canoes, bedecked in colorful feather plumes, each large enough to carry fifty or sixty men. Runners kept pace on the bank as the boats rowed downstream. As they paddled, the crews chanted and sang to keep their rhythm: taunts toward the bearded Spaniards, praise for their great chief. Despite the chaotic waters produced by that summer's flood, Quigualtam's navy slipped deftly over the river's surface, flanking and back-paddling, circling and splitting the little Spanish fleet. When they attacked, the Indigenous soldiers leaped into the river and swam to the Spaniards, overturning their boats. The heavy weight of their armor dragged some of the Europeans to their death. One man was pulled aboard an enemy canoe; his final fate has been lost to history.

All told, the Spaniards were under attack for nearly three days—first from Quigualtam's forces, and then, once they passed beyond his territory, from other nations downstream. As they approached the delta, the river turned relievingly quiet, except one small group near the river's mouth. (These appeared to the Spaniards to be a different people than the other nations they'd met—taller, with darker skin—though equally aggressive.) After more than two weeks, the last three hundred survivors of Soto's ill-fated expedition escaped the Mississippi Valley. Even then, they could see the marker of the river they'd traveled: a muddy plume extended far out to sea.

Today these men are known as the discoverers of the Mississippi River, which they first crossed in 1541. A mural depicting this accomplishment hangs in the U.S. Capitol. (The honors don't end there: towns and counties from Florida to Nebraska bear Soto's name, along with lakes and waterfalls and forests, bridges and schools and warships, and even a now-defunct line of Chrysler automobiles.) This is ironic, and not just in the typical fashion of colonialism—that tired idea that a white man can discover something that has already been used as a watery highway for thousands of years. Nor because another conquistador had crossed into the Mississippi's watershed a year before, far up the Red River. Rather, to call these men the finders of a river is to misinterpret their mission. Soto did not care about rivers at all.

The written record of the moment of discovery is revealing. One of Soto's soldiers, an anonymous Portuguese knight now known to historians as the Gentleman of Elvas, does little more than describe its breadth. His prose is sober: "A man standing on the shore could not be told, whether he were a man or something else, from the other side. The stream was swift, and very deep; the water, always flowing turbidly, brought along from above many trees and timber, driven onward by its force."

For Elvas, and for Soto, the river was simply an obstacle to cross. They did not record its coordinates, and if they made any maps, these have been lost. Perhaps it's no surprise, then, that it took another 130 years for a white man to return.

René-Robert Cavalier, Sieur de La Salle, began his rise in the 1660s, an ambitious young Frenchman set loose on a new continent. Born in France, he renounced an inheritance and pledged to a Jesuit seminary; then he renounced the Jesuits, too, casting off for the new colony of Canada with just a small stipend from his family. From a patch of wild forest, he built a fort and a fortune, becoming a key player in the northeastern fur trade. But that success was not enough.

The French had heard stories of some large river that flowed south, into an ocean, one that might offer passage to the Pacific—linking the so-called New World to the riches of the Orient. La Salle set out south from Montreal in search of this river in 1669. His party of fellow explorers soon abandoned him, convinced the quest was doomed. They were wrong: La Salle survived a harsh winter in the Ohio Valley, during which he may have become the first white man to see its river.

It's hard to know just what La Salle was up to. For the next few years, he wrote no letters, and no journal from this journey survived; the only records we have are the rumors shared by other backwoodsmen. Their reports suggest La Salle had learned the ways of the northern forests—the delicate arts of tribal diplomacy, the science of trapping and gathering food.

As it turned out, other Frenchmen beat La Salle to the Mississippi River. In 1673, departing from the headwaters of the Wisconsin, Father Jacques Marquette and the fur trader Louis Joliet, along with five other explorers, descended as far as the mouth of the Arkansas. That was far enough to determine that the river did not bear west toward China, as they had hoped—and far enough to hear rumors that the tribes to the south were armed and warlike. Despite that threat, La Salle figured this southern river would provide a smoother path for shipping furs to Europe than the often-icebound St. Lawrence. So in 1677, after returning to Canada, he sailed to Paris, where he sweet-talked the king into granting him a five-year license over the fur trade across the entire Mississippi River Basin.

La Salle burned through the first few years of his license just trying to get to the river. As his crew threaded up the St. Lawrence from Montreal, then through the Great Lakes, they weathered shipwrecks and insurrections and fevers and attacks by Iroquois soldiers. La Salle built several forts, one of which was called Crèvecoeur, "heartbreak." As if to fulfill its name, the fort was burned by more mutineers. Despite the challenges, La Salle vastly expanded the French knowledge of the continent. Under his instruction, two of his men turned north up the Mississippi, eventually reaching the river's sole waterfall.

La Salle's final descent of the Mississippi River commenced in January 1682, when his crew—twenty-three Frenchmen and thirty-one Indigenous laborers, including ten women and three children—dragged their canoes across the swampy ground at the edge of Lake Michigan. They found that the headwaters of the Illinois River were choked with ice, so they built sleds to lug their canoes a few dozen miles farther. Once they passed the ice, travel turned easy. Before February was out, the party was nearly five hundred miles south, beyond the mouth of the Ohio. On April 6, 1682, after feasting with friendly Indigenous villagers—and fighting off attacks by those who were not so friendly—the expedition reached the head of passes, where the river splits into three separate routes to the sea. They planted a cross in the mud, shot rifles, and sang songs—producing a great, noisy pageant to claim this river as a possession of their king.

I feel an uneasy envy toward La Salle. What he got to see! That ancient river can only be a dream to us today, wild nature as it appeared before the depredations of the white man. Cypress forests stretched intact across the bottomlands; the river ran free, undiked and undammed. But it's a mistake to believe that La Salle observed an untouched Eden. People had lived along this river for many millennia, shaping it to their needs. Much of the river valley had been converted to cropland centuries earlier; Indigenous people used fire to burn pathways through the forests. Other European immigrants had preceded the Frenchman, though these were different species entirely. Fig and peach trees sprouted within the river valley, perhaps carried here down trade routes that linked the river to the British colonies on the Atlantic coast. Chickens clucking in the river's villages may have been brought from the Southwest. Soto and his men had left behind hogs, which had become incorporated into local diets.

The biggest change, though, was human. Soto had found along the Lower Mississippi a center of wealth and power. La Salle encountered almost no one: a few Quapaw villages in the Arkansas lowlands, plus some Chickasaw hunters whom his men captured atop the bluffs near modern-day Memphis. In the Yazoo Basin, once the domain of Quigualtam, the expedition went entirely unthreatened, passing just a few small villages near the mouth of the region's namesake river. A credible estimate puts the population of the lower river valley at around a half-million during Soto's journey. By La Salle's era, 150 years later, the population had dropped to fifteen thousand.

"The collapse of the Mississippi valley chiefdoms and the virtual abandonment of much of the central river valley meander zone is one of the greatest historic puzzles of the American South," writes the historical anthropologist Robbie Ethridge. "We still do not have a good sense of why these chiefdoms fell, or about where the people went afterward." It was not just the Mississippi River that was suffering. The Ohio, too, was emptying throughout the seventeenth century.

One reason may have been the instability of the era's politics. The death of a chief could send a village into chaos. It's likely that Soto contributed too. His army laid waste to entire villages and ransacked food supplies. They sowed chaos in a region that was already marked by warfare. His men carried unfamiliar pathogens, too. And the lingering European presence to the north and east, beyond the watershed, would eventually bring more violence: a century after Soto's tour, Iroquois soldiers armed with European weapons descended the Ohio from their homeland in New York. To the south, the English ventured into the eastern edge of the Lower Mississippi Valley and set up camp among the Chickasaw. So here, too, rifles contributed to the turmoil.

Many tribes had a long tradition of seizing wartime captives and forcing them into labor or granting them as gifts or holding them for ransom—or, in more recent decades, in a telling change, adopting them into kin groups to replace family members lost to the epidemics.

Supplied with new weapons, tribes shifted this practice. Armed parties captured hundreds of women and children at a time, then turned them over to European slavers. Perhaps the deal was just too good: in an exchange for a single captive, a hunter could receive a horse, a suit of clothing, and a hatchet, plus a gun and ammunition. The first French explorers recorded stories of Chickasaw raiders storming into the river's bottomlands. They sacked villages, killing the men—who were armed with only bows and arrows—and seizing the women and children. Over the years, the British marched thousands of enslaved people out of the Mississippi Valley, which sent new waves of warfare spilling across the South.

Scholars describe the river in this era as a part of the "Mississippian shatter zone." It's a zone in time as much as in space: the chaos and instability lasted from the late sixteenth century through the early eighteenth. You can think of the river valley that La Salle found as less an untamed wilderness than a postapocalyptic landscape. Former farm fields were going feral. Villages sat empty and abandoned, their ceremonial mounds left to crumble in the wind and rain.

The old world was not gone entirely. Far downriver, just before La Salle's men reached the delta, they visited a Taensa chief who sat on a couch flanked by three wives and sixty white-cloaked elders. Across from his house stood a temple where two priests kept a sacred fire burning. The complex was surrounded by pikes that featured the impaled heads of conquered foes. There were perhaps nine Taensa villages when La Salle arrived on the southern river. Within two decades, as the shattering continued, the entire nation would consist of forty cabins.

From the point of view of European royalty, this muddy expanse was a backwater within a backwater: a little-known region tucked at the bottom of a continent that was an ocean away from the seats of power. When Louis XIV learned that La Salle reached the mouth of the Mississippi River, he declared the accomplishment "quite useless." Nonetheless, La Salle's offer to pay back the king for any losses won him permission to launch a colony near the river's mouth. By 1684, the veteran explorer was sailing west from France once more, headed for the river he'd claimed.

This proved a difficult journey. In its early days, La Salle fell into a fever; while he writhed in bed, Spanish pirates seized one of his four ships. When the fleet paused in the Caribbean to recover, the soldiers "plunged into every kind of debauchery and intemperance so common in those parts," as one of the accompanying priests complained. Some sailors died; others deserted. With what remained of his crew, La Salle departed Hispaniola in late November. Quickly the boats were surrounded by turbid water, then buffeted by winds.

Finally, after days of difficult sailing, La Salle spied land on the horizon. The winds were favorable. How could he have felt anything but relief? With his trusted scouts, La Salle headed to shore.

They did not linger. They had found nothing promising, just low-lying grasses and marsh reeds, strewn with massive fallen trees. The fleet sailed on, still looking for its river.

Eventually La Salle ran aground in Texas. He wandered the edge of the gulf for years, trailing behind him an ill-prepared band of civilians who were supposed to be his colony's seed. Starving and delirious, their numbers dwindling, they grew weary of the hunt. In 1687 a few of the colonists turned mutinous and shot La Salle in the head. They stripped his body and left it naked in the brush. Soon enough they turned on one another, too. Almost everyone died.

The cause of La Salle's confusion remains debated, but notes in his logbooks suggest that he had accurately calculated the coordinates of his destination—that he had arrived safely back at the river's mouth. Here's one likely possibility, then: when he looked at the swampy landscape, he recognized nothing. He could find no river amid the marsh. The famed historian Francis Parkman called this La Salle's "fatal error."

I'm sympathetic. More than three centuries later, on a canoe trip from New Orleans to the Gulf of Mexico, I spent two nights camped on a small island at the river's edge. Looking back toward the continent, I saw grass and cane, humped together, hissing in the breeze. I saw no river. There *is* no river. It has already divided itself into hundreds of channels, tiny trails of water through the marsh.

The problem for La Salle and his pursuers was that Europe simply had no landscape like this one. There is nothing like it in the world. Deltas are shaped by rivers and ocean in a back-and-forth interplay, the piled-up mud fighting against the destruction of the tides and the ocean waves. Sheltered within the gulf, the Mississippi's delta is river-dominated, more so than any other major delta in the world. This makes its shoreline less regular, far more pockmarked with marshes and bays. The rivers the Europeans knew were more like trickles of water running atop stretches of dry land. Here was an endless expanse of mud.

It could not have helped that the maps of the day were crude renderings, featuring rivers and bays that were based mostly on hearsay and tall tales. Spanish ships had been exploring the gulf for more than two centuries, never noting the river itself. In the years after La Salle's failure, they scoured the sea for the Frenchman, determined to run their rival from a gulf they'd claimed. Their explorations of the coast were more intense than ever before—but not only did the Spaniards fail to find La Salle, they failed to find the river. And so, for another decade and a half, the written record of the watershed went blank again.

The old Bayou Choupic portage (Orleans Parish, Louisiana).

2

Tomahawk Claims

A changing of the guard

We can make at best a rough map of the Mississippi watershed at the tail end of the seventeenth century, as La Salle and his colonists fumbled across the coast. Even choosing what labels to apply is difficult, because the word *tribe* as we use it today is a modern invention. Sets of villages might come together into a loosely knit nation, but each village retained its own leader and made independent decisions. Nor was the Indigenous map settled and static. Boundaries were hazy, "felt rather than seen," as historian Pekka Hämäläinen puts it, and often centered around a sense of connection with a specific place or a stretch of river—and they could shift easily. Small-scale wars broke out in the hunting lands that served as border zones.

Sometimes people hauled themselves across the continent. The Ojibwe, for example, who live near the river's headwaters, had abandoned the eastern oceans more than a thousand years earlier, after a prophet warned the People that they must move west. The Quapaw, meanwhile, got their name because of a trip gone awry. Several nations were making their way west through the Ohio Valley; when they arrived on the

Mississippi River, they found the water wrapped in fog. They braided a rope of grapevine and began to ferry themselves across. The first few parties succeeded; on the far side of the river, they continued to march west, eventually settling along the Missouri. But as the last group reached the water, the rope broke. So these people floated south and settled in the Arkansas lowlands. They became the Quapaw—the downstream people.*

Despite these difficulties, a sketch is in order, if only to clarify the diversity of this place. The Ojibwe had finished their journey in a territory that straddles a triple divide, featuring the headwaters of rivers that run into three different oceans—including the two-thousand-mile route south, down the great river, into the Gulf of Mexico. Not far below lay Dakota country, which was centered around the Bdote, the coming together of the Minnesota and Mississippi rivers. As the eighteenth century began, the Dakota spent their summers in villages along the big river, where they gardened and fished. In late fall, they ascended the smaller tributaries to hunt. When the first French fur traders arrived in the northern watershed in 1679, they stopped at Bdé Wakháŋ—Spirit Lake, the headwaters of the Rum River, a Mississippi tributary—and watched as the Ojibwe and Dakota negotiated some kind of an agreement.

Some Dakota people were spreading west then, along the Missouri River, whose basin was occupied by other Siouan-speaking communities: Hidatsa and Mandan, Kaw and Osage, Omaha and Ponca, Iowa and Missouria. These groups may have all originated along the Mississippi, though their journeys west occurred so long ago that their languages have diverged and are no longer all mutually understood.

To the south, near where the Missouri and Mississippi rivers meet, a dozen Inoca communities were scattered along the nearby tributaries, growing corn and beans and melons. The Ojibwe called these southern

* One of the boat ramps in Mississippi, near John Ruskey's headquarters, is called Quapaw Landing. Because he used this landing frequently, and liked the name, Ruskey decided to call his business the Quapaw Canoe Company. Only later did he learn the word's derivation. But he very much considers himself a "downstream person," one who prefers, when possible, not to fight the flow.

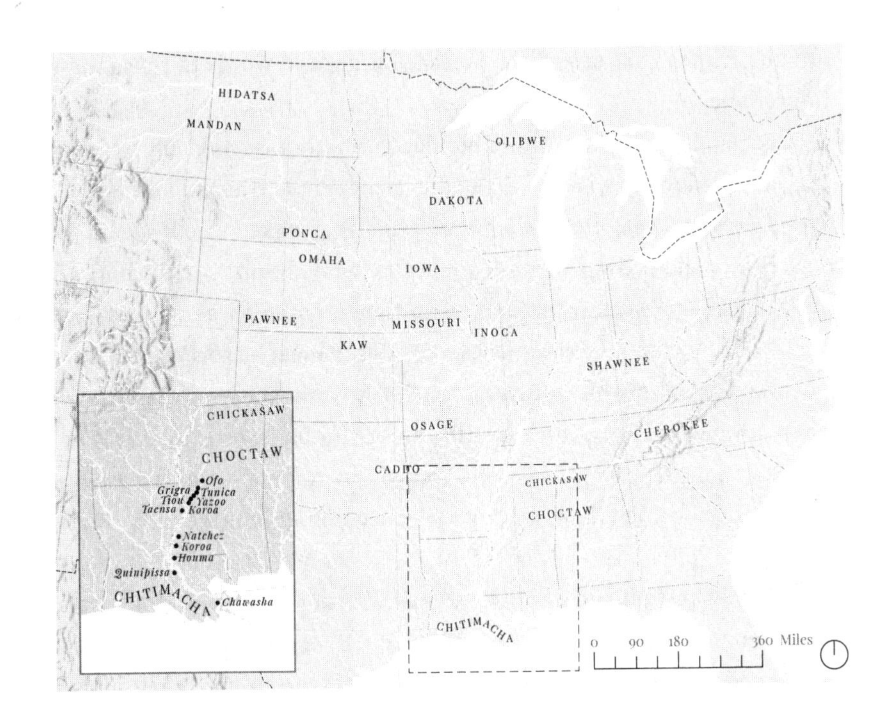

Indigenous villages and territories in the late seventeenth century.

neighbors something close to "Illinois," which is the name the French adopted. Other nations speaking similar Algonquin languages spread east along the Ohio River. Along the Tennessee, the Cherokee spoke an Iroquoian tongue. The forests along the Lower Mississippi's big western tributaries—the Red, the Arkansas, the Ouachita—were Caddo country.

The big powers of the South, like the Choctaw and Chickasaw, tended to live beyond the river valley, in the surrounding uplands, though they had a few villages within its forests, which served as key hunting land. The lower river, the delta especially, in the wake of all the shattering, had become a diverse place, home to the Quapaw and the Yazoo and Griga and Tiou and Natchez, the Koroa and Ofo, the Taensa and Tunica and Houma. *Les petites nations*, "the small nations," the French called these villages. The Choctaw knew one site in the delta, or perhaps the delta as a whole, as Bulbancha, "the Place of Many Languages." Wandering traders developed a lingua franca that mixed the varied sounds and words of these many different people.

Food along this river was abundant, whether hunted or fished or farmed. The lower river was especially rich. People trapped fish in backwater lakes and collected persimmons that they baked into loaves. They wore the dried heads of deer as camouflage when hunting in the bottomlands. The French reported that when they crossed paths with a Choctaw hunting party, they were sure to be offered meat—a sort of generosity that was typical of the culture. At each Choctaw house, a visitor could find a pot of *tan fula*, a thick hominy stew, simmering, ready to eat as a welcoming meal. When fish were dragged from ponds, any excess was sent to neighboring villages. At ritual dances, chiefs ensured that everyone was fed. This was true not just of the Choctaw. Across the Southeast, the political system—along with the social structure and economy—was built around an idea of reciprocal obligation. Every relationship, between people and their neighbors, between people and the animals they hunted, between people and the animals that might hunt them, was considered personal and intimate, built on a presumption of respect. For some Indigenous people, even an enemy could, through the right ceremony, become future kin.

By the tail end of the seventeenth century, Louis XIV decided he wanted this river after all. The English traders who had traveled west from the Carolinas to arm the Chickasaw were collecting deerskins and human slaves from land that the French had supposedly claimed; Spanish sailors were wandering the gulf, brushing against the delta. In the meantime, the French were making their first attempts to settle the river valley. La Salle had granted one of his lieutenants land near the mouth of the Arkansas River, where he set up a trading post among the Quapaw, a place to gather beaver furs. French priests had settled along the Illinois River, among the Inoca. So Louis decided to send an explorer to secure the watershed.

Pierre Le Moyne, Sieur d'Iberville, was almost forty years old, and had distinguished himself in Canada as a capable soldier and a trustworthy navigator—one who might actually be able to find the river. He was told to select a "good site" near the Mississippi's mouth that could "be defended with a few men, and block entry to the river by other nations." According to one member of his crew, the men arrived at a river mouth that consisted of "nothing more than two narrow strips of land, about a musket shot in width." As they proceeded up the Mississippi, they noted a vast expanse of marshland beyond those humps of land. The place sparkled with water and teemed with birds. The few trees were tangled in vines. They had rediscovered the delta.

By 1700, the French were at work amid this soggy land, fifty miles above the river's mouth. They burned plants, chopped logs, dug mud, then erected a half-dozen palm-roofed shacks and encircled the camp with cannons. This first French "fort" was troubled by the river's strange geography. The strips of land that run along the river's final miles dangle into the Gulf of Mexico such that the Mississippi's mouth lies nearly a hundred miles south of the coastline to the east, beyond the delta. That means that if you are departing a seaside village in Florida or Alabama and want to reach, say, the Quapaw villages near the Arkansas River—or Inoca territory, or the fur-trading posts far to the north, near the Mississippi's headwaters—then to enter the river by its mouth is an arduous trek. You have to paddle thirty miles south into the ocean before curling back north.

Nor did it help that the mouth was often blocked by shoals—which a canoe might pass, but a deep-draft ship could not. This terrain was miserable, anyway. The only animals to be found were gnats and mosquitoes; game was all but nonexistent. A visiting priest noted that due to flooding "the men spent four months in the water." Simply leaving your cabin might require wading through water halfway up your legs. The garden was swamped, too, and beset by "great numbers of black snakes" that ate the men's produce down to its roots. The place was abandoned before the decade was out, and the French did their best to secure their river from the towns they built far to the east, on the more solid ground along what is today the coast of Mississippi and Alabama.

The French had more success to the north, in the "Illinois Country," where several missions established along the Mississippi at the dawn of the eighteenth century grew into substantial wheat-growing villages. In Louisiana, meanwhile, a sequence of settlers attempted, mostly unsuccessfully, to discern some use for all the mud. Perhaps there were minerals. A businessman named Antoine Crozat hoped so, and in 1712 he leased the whole of Louisiana from the king. Over the next few years, colonists established several lasting outposts—one atop the bluffs now known as Natchez, and another up the Red, at the Caddo village of Natchitoches—but Crozat soon abandoned his claim. To increase the value of the land, and to make up for some of his losses, Crozat suggested that tobacco, an increasingly popular vice in Europe, might be easily grown in this ground.

This was, if not a lie, then at least a wild speculation, but it proved enticing to John Law. Known as a mathematical genius, a devoted gambler, and ultimately one of the greatest swindlers in history, Law had already launched a bank that printed the first paper currency in France. Now he sought and received exclusive trading rights for France's North American territories. The Company of the West, as he called his new venture, got to appoint the governor and other officers, and promised to deliver six thousand settlers, along with another three thousand enslaved Africans. Law's plan was to corner the tobacco trade, one part of a complex scheme to prop up the drooping French economy. So under his watch, plantation agriculture arrived on the Mississippi River: massive farms all featuring a single plant species that, once harvested, would be shipped across an ocean. Unlike Indigenous agronomy, there was nothing personal in this approach. Plants were commodities, meant to be rendered into money—and to turn a profit, these farms depended on the work of the enslaved.

Company officials had selected Iberville's younger brother, Jean-Baptiste Le Moyne, Sieur de Bienville, as the *commandant-général* of the fledging colony. Needing to establish a headquarters, Bienville settled on a site a hundred miles above the Mississippi's mouth. It was a place long used by Indigenous people, who knew that rather than bothering with the long paddle south, the easier entry point onto this river was the estuarine bay of Lake Pontchartrain. If you picked one of the small streams that fringe its southern edge, then paddled against the nearly nonexistent current into the cypress forests, you'd eventually reach the back side of the Mississippi's natural levee. After a short portage—a mile or two—you could depart northward toward nearly half the continent. Iberville and Bienville had long ago noted that the portage from the bayou to the big river had been established as a "rather good road." In 1718, at the road's riverside terminus, Bienville ceremonially cut a few pieces of cane. Then he turned the work over to a team of laborers.

When one colonist visited the site soon afterward, the future city of New Orleans consisted of one tiny bare patch carved out of the seemingly endless expanse of cypress forest. The sole building was a hut for Bienville, its roof thatched with palmetto leaves, in the Indigenous style. The next year, while the village's first buildings were under construction, the river swelled over its banks. So Bienville put laborers—including men kidnapped from the Senegambia region and shipped to the French colony, some of the first enslaved Africans along this river—to work erecting the Mississippi River's first levee.

The Company of the West had expanded quickly. Law acquired trade rights in Africa and the Caribbean, bringing in enough profits to make many of his investors into wealthy men. But the conditions in Louisiana proved troubling. Fever was rampant. Food was short. Hurricanes barreled in from the gulf. Law's backers, sniffing out the troubles, began to divest. By the time the river's first levee was in place in 1720, the company was already undergoing corporate restructuring. Its complete collapse the next year is one of the most famous burst bubbles in financial history.

Louisiana was in shambles. Within a few years, half the colony's population—many of them army deserters and petty criminals Law had swept up from France's prisons in a desperate effort to find willing settlers—were gone. Some had fled back to Europe. Many were dead.

By now, the whole watershed was shattering. Only the Ohio River was free of European encroachment—the imperial powers saw its valley as too remote for settlement—and so its headwaters became a kind of refuge, at least for a few years. Still, the distance was not enough to stop the spread of epidemics, or the trade in rum, which the French and British sometimes offered in exchange for furs. There was no real refuge when the continent was undergoing so much change.

The growth of the fur industry had disrupted societies along the upper river, too, stoking new rivalries. But the lower river was the site of the worst of the warfare. Before the Company of the West was established, Louisiana had been home to just a few hundred wandering European hunters. Now, even after the collapse, thousands of white men and women were occupying riverside land. The one great power along the river were the Natchez, who may have been the descendants of Quigualtam's followers. Their villages perched on the river's southernmost bluffs, which stand two hundred feet above the river—terrain that was more familiar for the European colonists than the delta's morass. This became the chosen site for the first efforts to plant tobacco. The Natchez chief, the Great Sun, watched for more than a decade as plows tore through the ground, ripping loose soils. Horses and pigs and cows trampled bottomlands, which were key sources of food. By 1729, the Great Sun had had enough.

That summer a battalion of Natchez soldiers entered a small fort that the French had built atop the bluffs. The men signaled friendly intentions, but once inside, they raised their rifles and executed the French soldiers at close range. On all the surrounding plantations, the patriarchs were shot or clubbed, then scalped and beheaded. Of the 150 white men who lived in the region, twenty escaped alive. (Twelve Natchez died in the fight.) The victors slaughtered the white men's cattle and cracked open their brandy, commencing a celebration while the fort burned to the ground. Then they set to work, building new defenses against the retribution they knew would come. Indeed, the French—or their Choctaw allies, really, who contributed the bulk of the thousand or so soldiers—soon launched a siege. After a month of tension, the Choctaw managed to negotiate the release of the 150 women and children whom the Natchez were holding hostage.

On the night after those negotiations, under the cover of darkness, the Natchez slipped down the bluffs, into the swamps across the river. They were hunted down over the next few years—sometimes killed upon capture, sometimes sold into slavery in the Caribbean. Some managed to escape into nearby tribes: as many as a hundred Natchez formed their own village among the Chickasaw, near the headwaters of the Yazoo River. In Cherokee and Muscogee towns, the Natchez became known as important spiritual leaders. At least we know what happened to this nation. Other groups, like the Quinipissa, simply disappeared from the colonial records, never to reappear.

After the war against the Natchez, Louisiana's government spent years assembling a massive force to dispatch against the British-allied Chickasaw. The thousand Frenchmen and five hundred Indigenous soldiers, mostly from the Great Lakes, constituted one of the largest armies yet assembled by a colonial power in North America. But after a few brief skirmishes in 1740, the two camps negotiated a tenuous peace. The French were reluctant to wipe out one of the region's last great confederations, as it might make their Choctaw allies less pliable. They needed these soldiers to fight on their behalf.

Despite their weak hold on the colony, the French kept expanding, turning their attention to the Ohio River by the 1750s. The British, too, had been eyeing this valley. So soldiers scurried through the woods. Forts arose.

Near sunrise on a May morning in 1754, a twenty-two-year-old lieutenant colonel named George Washington, leading a party of Virginia militiamen, found himself on a precipice in the Pennsylvania Alleghenies, overlooking a small French encampment tucked into a glen. No one knows who shot first, but the French were routed, and the result was a global war that lasted almost a decade.

In North America alone, the Seven Years' War featured skirmishes from the Great Lakes to Cuba. Few official battles were fought in the Mississippi watershed, though the local tribes waged a proxy war on behalf of their chosen allies, and the end of the conflict resulted in a major upheaval: the colonial maps were entirely rewritten. In 1763 France ceded its North American holdings. Almost everything east of the Mississippi went to Britain, while everything to the west, plus the city of New Orleans, went to Spain, as recompense for its allyship. Less than a hundred years after La Salle claimed this watershed, French rule was over.

For the Indigenous people who lived along the river, this war was just one more episode in the long years of shattering. This fact became clear to me one summer when I was hiking along a bluff in Minnesota. At the edge of the trail, I saw divots in the soil, as if deer had dug out beds for a nap. Their true purpose was somewhat less wholesome: archaeologists believe these are the remnants of rifle pits dug by Ojibwe soldiers in 1768. As the story goes, the men lay in wait, then ambushed a Dakota war party traveling the river with kidnapped Ojibwe women. The battle lasted two days and ended in Ojibwe victory, cementing their claim to the upper river. Unlike George Washington's battle in the Alleghenies, this skirmish went unrecorded in the colonial records. The details were passed down through the generations by the Ojibwe until a hundred years later a historian finally put them down in his book.

In the eyes of the British, the whole of the watershed now became a forbidden frontier. King George III drew an invisible line across the top of the Appalachian Mountains, declaring the region beyond closed to white settlement. The French may have lost a war, but the Indigenous people across the Mississippi River's watershed had not. Seeking stability, the British crown decided to recognize their claims to the land.

This did not stop the settlers.

Back when Hernando de Soto arrived on the river, he had had no sense that he could own the surrounding land. That privilege lay with the king alone. Early European empires tried to plan a colony, then sent out the settlers to make their vision real. Of course, amid the vastness of the Mississippi's watershed, the monarchs did not always have control; hunters and traders could claim a homestead, and their king would never know. Still, the British settlers who floated down the Ohio tributaries in the wake of the Seven Years' War represented something new: not a slow dribble but a great tide of humanity large enough to upend imperial plans.

George Washington joined the land rush, ordering his personal surveyor to search for viable acreage. (If confronted by royal officials, the surveyor was instructed to claim he was simply out on a hunt.) But rich men like Washington weren't considered the problem. The crown was more concerned with, as one governor put it, the "overflowing Scum of the Empire." The procession into Kentucky was at times a parade of barefoot settlers; in lieu of cash, the economy ran on trades in tobacco, deerskins, cows, and calves. These settlers set up camp in the bottomlands, atop ground their Indigenous predecessors had long before cleared for growing corn. The royal soldiers made some effort to stanch the flow, burning down illegal encampments near the Ohio headwaters in 1767. Within months, the squatters returned in bigger numbers.

One key idea fueled this scummy migration: these settlers believed that by improving the land, it became theirs alone. Nature, in other words, was a blank slate, a divine gift, awaiting its conversion to private hands. The Mississippi was a particularly enticing blank, beginning at "a source unknown," as an article in *Freeman's Journal* put it in 1782, and passing "savage groves, as yet uninvestigated by the traveller, unsung by the poet, or unmeasured by the chains of the geometrician."

The life of George Rogers Clark offers a suitable synopsis of the era. Red-headed, standing six feet tall, Clark crossed the line in 1772. He was a nineteen-year-old who had quit formal school, training instead under his grandfather to become a surveyor. He carried a rifle and a blanket roll and a copy of Euclid's *Elements* as he traveled three hundred miles down the Ohio River, often described in this era as the world's most beautiful waterway. He passed from craggy mountain caverns into soft hills into the wide valleys of Kentucky. When he returned to the mountains near the headwaters, he picked a plot of land tucked along the mouth of a little creek. He slashed marks in trees—a "tomahawk claim"—denoting that this was now his property. He cleared the woods and built a cabin. He raised corn. Then he went to war.

First Clark fought the Shawnee, a people who had recently returned to their ancestral Ohio homelands after years of wandering. Theirs was not an easy arrival: a series of small skirmishes with the new frontiersmen set off a series of massacres, which set off more retaliatory massacres.

The fighting culminated in a bloody riverside battle in 1774, a victory for the Virginia colonial militia. A few years after that battle, Clark was back to soldiery, now as a rebel against the British crown. He led a contingent west to the Mississippi River, where, on July 4, 1778, without firing a shot, he seized Kaskaskia. The town had been founded by the French as a mission and fur-trading post; eight decades later French remained the language of choice, and the residents had little loyalty to the British crown. (To the south, meanwhile, the Spaniards were using the Mississippi River to send munitions up to the rebels.) Clark spent the next several years securing the territory, in a new war that overlapped with the old. Clark made a point of razing Shawnee villages as he marched through the Ohio Valley, destroying hundreds of acres of cornfields just before harvest.

This anti-Indigenous violence continued along the Ohio River for decades after Britain conceded. But Clark moved on to more domestic matters. He laid out the streetscape for a thousand-acre village near a key set of rapids on the Ohio River, naming the place for himself: Clarksville. Almost immediately the federal government repeated the sin of the British, declaring that *it* owned this land. Worried that developing it might set off more wars against the Shawnee and other tribes, the new government deemed Clarksville's twenty-three residents to be squatters with no valid claim to this ground.

Many easterners thought the Mississippi's watershed was worthless anyway; the mountains were a wall that separated two territories that were destined to become two separate nations. This view prevailed in 1785, when the United States began to negotiate its first treaties with Spain and proved willing to give up the right to navigate the river in exchange for, among other trading rights, access to Canadian fisheries in Newfoundland. The frontier became once more a rebellious hotbed; Clark himself wrote to the Spanish government in 1788, proposing that he establish a town near the mouth of the Yazoo River under its flag.

Still, the watershed kept on drawing its astounding waves of settlers: more than one hundred thousand people clambered over the mountains in the last three decades of the eighteenth century. These new arrivals felled forests to build homes and fences and house-warming fires—so many trees that before the century was out, travelers were reporting that the banks of the Ohio had been entirely shorn.

Despite the Spanish ban on American traffic, these farmers rode flatboats downriver and unloaded at the levee in New Orleans. The first recorded voyages launched from Pittsburgh in 1782—ten boats carrying three hundred tons of flour—but the U.S. general who described this business spoke as if it had been well established for years. By 1795, the Spaniards conceded the point and officially opened the river to U.S. traffic. They'd never really managed to securely govern this place, and five years later they gave up trying, ceding Louisiana back to France. In his registry of goods distributed to the local tribes, the departing lieutenant governor of Upper Louisiana left a succinct summary of the worth of the place: "The devil take it all."

Thomas Jefferson had long disagreed with the notion that the Mississippi watershed was beyond the reach of America. Years earlier, as secretary of state, he'd suggested that New Orleans lay not on land but on "a streight of the sea." (It is "only here and there, in spots and slips, that the land rises above the level of the water in times of inundation," he wrote.) This little sliver of foreign territory was all that separated Kentucky farmers from the ocean, so slight that the geography alone implied that Americans should be allowed to pass through. Now that he was president, Jefferson sent a diplomat to Paris on a mission to purchase New Orleans and secure this passage. The French, as it turned out, were bogged down by war and willing to sell the entire territory. So Jefferson purchased half of one of the world's great watersheds for the bargain price of $15 million, roughly four cents per acre.

The United States of America doubled in size. The Mississippi River was no longer its western limit but its heart. The whole watershed was at last claimed by one single imperial power, and its great renovation was set to begin.

Winterville Mounds (Washington County, Mississippi).

3

Cosmic River

Reading the stories of the earthworks

Among the expeditions that Jefferson funded to explore this territory, one has been mostly forgotten. The president asked a bookish Natchez-based planter named William Dunbar and a Philadelphia chemist, George Hunter—both Scottish-born, both casual adventurers—to head west to the Red and Arkansas rivers. Thanks to war among the Osage and the presence of Spanish soldiers, everyone got cold feet. (The route would later be assigned to several other explorers, including Zebulon Pike.) So the men ascended the Ouachita River instead, a six-hundred-mile tributary of the Red that begins in Arkansas and pours through the swampy lands of northern Louisiana.

In the late winter of 1804, the crew found empty shacks and cabins near a set of hot springs in the mountains, a pleasant place to pause for Christmas. Tiny crustaceans lived in the scalding hot water, which, when cooled and sipped, tasted something like spicewood tea. On the way home, in the black-soiled lowlands fifteen miles west of the Mississippi River, the explorers paused to document a spectacular landmark: on a spit of land where three creeks meet, five carefully arranged earthen pyramids rose above the jungle. These may have been "a temple for the adoration of the Supreme being," as Dunbar wrote in his notes, a sign of some long-lost civilization. Now, the mounds were a backwoods landmark that was nearly inaccessible through the thick growth of river cane. The surrounding region was occupied by a few cattle drivers, living lives that Dunbar considered barely civilized.

Across North America, Indigenous people carefully arranged soils into monumental constructions: pyramids and cones and hillocks, embankments and enclosures, silhouetted effigies of animals and humans and spiritual beings. *Mound* is the familiar word for such sites, used by writers Indigenous and non-, but perhaps it is better to think of them as *earthworks,* a juxtaposition of "grounded *earth* [and] dynamic *works*" that, as the Indigenous scholar Chadwick Allen notes, emphasizes the labor and thought required. Thousands of these earthworks have been documented across North America, from Oregon to Texas to Florida, almost always close to rivers. Nowhere are they in greater concentration than along the Mississippi and its tributaries.

The five pyramids that Dunbar and Hunter discovered included the largest earthwork yet recorded by settlers. It covered an acre of land, rising in a steplike pattern through three tiers until it reached eighty feet. A wall of soil, ten feet tall and ten feet wide—taller than almost any levee in this era—ran along the edges of the site. But while Jefferson noted these mounds in a brief to Congress, no one made much of an effort to preserve or even study this place. By the time Mark Twain visited in 1882, the town of Troyville had been platted within the earthen wall. Houses had been installed at the bases of the earthworks. The river was flooding, and Twain found that the mounds, the only ground that rose above the water, were crowded with sickly hogs and mules and cows. Already these monuments were beginning to lose their shape: during the Civil War, soldiers had dug out a rifle pit in the largest. Later, as the town of Troyville grew, its citizens treated the constructions' soils as something to mine so that they could level yards and fields.

The state of Louisiana treated this ancient monument as a convenient supply of fill, too, when it built a bridge across a nearby river in 1931.

Dismantling the earthworks turned out to be difficult, requiring a full month of around-the-clock shifts—horses and scrapers, laborers wielding picks, steam-powered shovelers—so as to break apart the clay that had been carefully laid thousands of years before. Eventually, though, the largest mound was demolished. The ancient dirt was piled up to support the bridge's western approach.

The next year an archaeologist visited, to salvage what he could. He concluded that the site must have been "the capital of an extensive province," and that these swampy bottomlands had once supported a much larger population than the sad little hamlet of Troyville. Perhaps he was biased against the place: the locals were suspicious, convinced he was seeking a treasure that had been buried by river pirates. Their demands for payment "finally became so unreasonable and so impossible to grant that the excavation of the great mound site had to be summarily stopped," the archaeologist wrote in his report.

I passed through the town a few years ago and found Jonesville, as it's called now, unshakably spooky. The roads are potholed; some of the downtown lots have been reduced to rotting piles of insulation and plywood, though it was unclear to me whether the precipitating disaster was a flood or a tornado or simply neglect. One of the old mounds had escaped destruction, perhaps because the town's first residents commandeered its bulk for their cemetery. A cypress tree now stands above the slanted gravestones. As I climbed its rise, a woman drifted past in the street below, barefoot and ghostly. She seemed to live in a separate world, not heeding my presence at all. A few blocks away, at the site of the great mound, the bottommost three feet of soil remain—though if not for the sign posted nearby, I would never have noticed. It's grown a cover of clover, and a Catholic church has been built on top, a midcentury A-frame, a strangely modern design for this town.

As I turned out of Jonesville, back onto the state highway, I noticed a small and lumpy pile of soil. A plastic bag was caught in its grass. An ironic sight, I figured: a landfill, our typical modern American earthwork. But a sign revealed the truth. This was a scaled-down replica, a belated attempt to conjure what had been destroyed.

The earthworks in Jonesville suggest a different Mississippi River story, one with a different beginning—another murky beginning, since we don't know just when the first people arrived along the Mississippi River. Evidence suggests that migrants passed along the Gulf of Mexico more than fourteen thousand years ago, before the last glaciers had fully receded, reaching as far as Florida. The first time people pushed into the alluvial valley, and whether they came from the north or the south, is a matter of debate, but the hemisphere's oldest formal open-air cemetery was established in the river's floodplain in Arkansas roughly twelve thousand years ago. Four thousand years later people in the Southeast built the first examples of an enduring form of spiritual architecture: earthen mounds.

Eventually, these mounds morphed from simple piles of shell and soil into something far more architectural. Mounds were built in sets that were arranged in circles, in some cases aligned with the rising and setting of the solstice sun. Over a thousand-year period that began approximately 5,500 years ago, more than a dozen such complexes were built in Louisiana and Mississippi, always on the high ground alongside the river's bayous. The grandest of these sites, now known as Watson Brake, features a ring of eleven mounds, some more than twenty-four feet tall, connected by a lower wall of earth. These sit atop a terrace that at the time of construction probably overlooked a clear-running side channel of the Ouachita River, near the edge of the Mississippi's vast floodplain. The mounds were perched, then, between dry forests and watery swamps.

Watson Brake arose in occasional bursts of construction over at least six hundred years. Centuries sometimes passed with no progress. To put that to scale: six hundred years ago, Christopher Columbus was not yet born. Whatever culture produced Watson Brake, it lasted longer than the settler experiment so far.

Archaeologists suggest that this site served as base camps for villagers who went hunting and fishing in the surrounding landscape—people who, so far as the evidence shows, were egalitarian, with no one group receiving any kind of royal treatment. But what did the architecture mean? We cannot know with any certainty. Nor can we know why, after a thousand years, all the sites that had been built were abandoned. The different groups seemed to break apart, reverting to an older way of living, scattered lightly across the land. As far as archaeologists are concerned, it's as if the people melted away into the wilderness, leaving no more record.

One potential culprit is the river itself. Due to long-term fluctuations in the climate, its rate of floods sometimes jumps upward. For two centuries, or five, or a thousand, floods will shudder through the river valley; then for equally long periods the river will drop into quietude. Watson Brake and its peers were built during one of these calmer periods. At the time when the mound sites emptied, floods were beginning to crash through the local bayous with renewed vigor—not unlike the crisis we face today.

People would have seen the old ruins. As they wandered the river valley—hunting deer, hooking fish, collecting pawpaws—perhaps they passed down legends that explained the mounds' meaning. Perhaps they regarded them with the same wonder that archaeologists do today. Either way, after forty generations—a thousand years!—people started building again. Once more, scattered at a half-dozen sites across the Mississippi's southern valley, clusters of small mounds arose. But this time, the world of mounds had a distinct center.

We call this central node Poverty Point, the name of a cotton plantation that a white man farmed in the nineteenth century—a name that may suggest the farm's lack of success. To the people who built the mounds, poverty must have been unthinkable. They had no interest in growing crops, not when there was so much to eat in the backwaters nearby. Poverty Point's chefs roasted cattails and lotus tubers in earthen ovens; people ate persimmons and pawpaws, acorns and nuts, deer and turtle and raccoon, plus lots and lots of fish.

Poverty Point sits just fifty miles as the crow flies from Watson Brake, atop Macon Ridge, a strip of "second bottom" uplands—a higher ridge of coarse sediment delivered by the earlier, braided river—that runs through the western alluvial valley in a wide, north-to-south band. The mounds were built on the eastern edge of the ridge, where bluffs drop away into the swamps. The largest, a wide, flat-topped platform, stands seventy-two feet tall, higher than even the largest of the modern levees. The dirt contained in this mound would fill thirty thousand dump trucks and was carefully selected from several nearby sites, creating a mixture of textures and colors. Travelers delivered goods to this place from up and down the big river: Galena rock from Iowa and Missouri, Pickwick chert from Tennessee, flint from Indiana. Red jasper pendants carved in the shape of pot-bellied owls were carried *away* from Poverty Point, to sites as far afield as Florida.

Why do all this? We will never know for certain. No one has found evidence of permanent home sites or human remains here, so many archaeologists believe that Poverty Point was not a village but a seasonal gathering place, where people came to make art and tools and to trade. Perhaps various groups joined to feast and intermarry and pile up earth—building, quite literally, a shared world.

Poverty Point, too, was abandoned after six centuries. Mound building continued, but at a reduced scale. A few centuries later a new kind of construction emerged far upriver, in the Ohio Valley, in a style with impressive elaboration: embankments that zigged and zagged and encircled the mounds, setting off a sacred zone. But the next great mounds to be built along the lower river would be the complex that Hunter and Dunbar found, with its strange tiered pyramid. These earthworks were not built until another thousand years after Poverty Point.

Through the centuries, new styles of mounds kept appearing; famously, the Upper Mississippi features effigies of panthers and bears and thunderbirds. On the Lower Mississippi, however, the custom remained steady and simple: mounds that featured flat tops, as at Poverty Point, arranged around plazas. More than anywhere else in the country, the people along the southern river—the place where mound building first flourished—seemed committed to upholding the earthwork tradition as it had first begun.

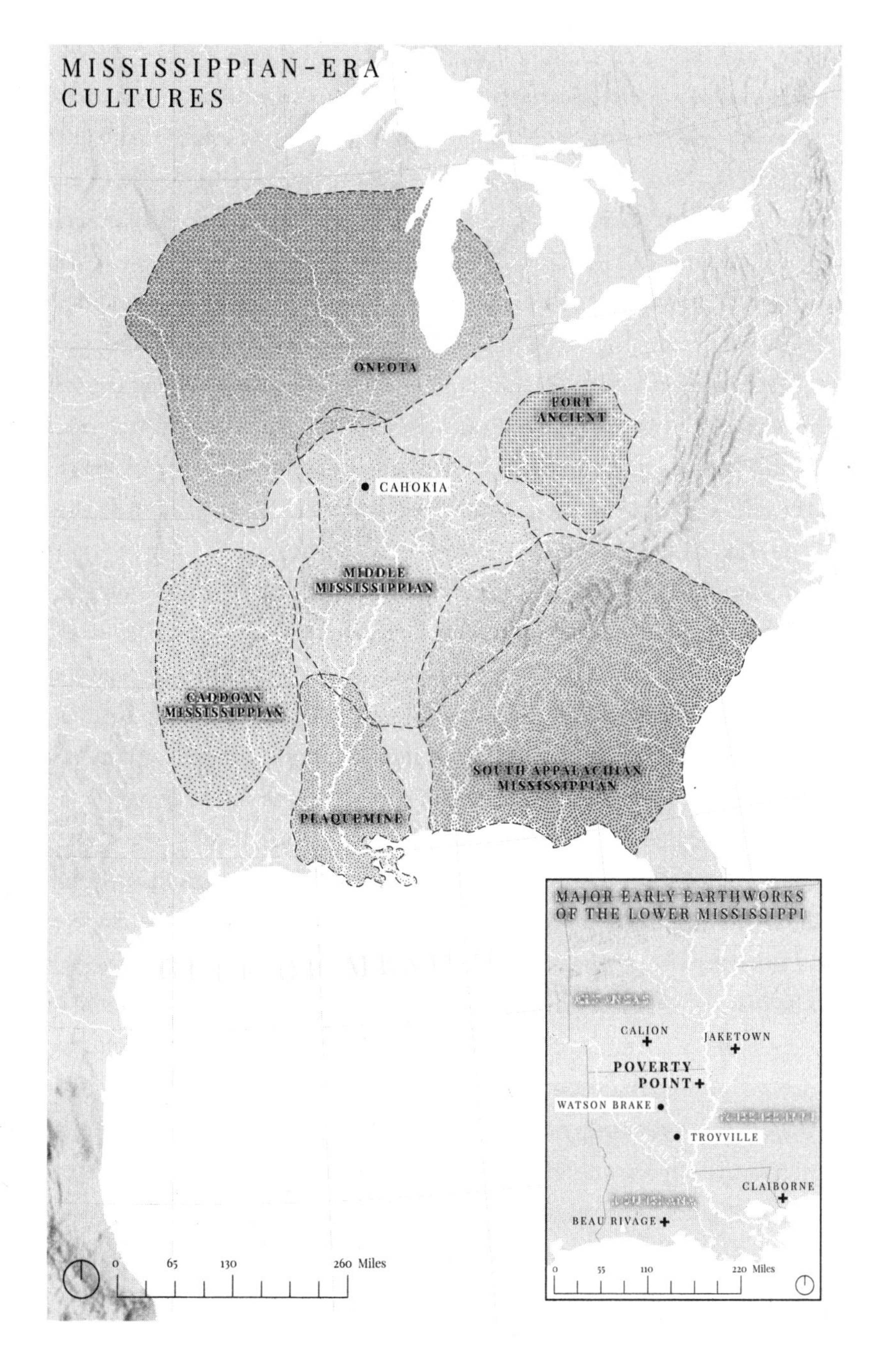

Major mound-building cultures and sites.

That is not to say that this style of building was limited to the Lower Mississippi. The most famous example of a flat-topped mound is in the American Bottom, the low swath of floodplain across the river from St. Louis. This valley, torn open by glacial meltwater, spans ten miles across at its widest. It's a striking place, as if a sliver of the Lower Mississippi's swamps were tucked into a notch in the Ozarks. The long horizontal lines of turnrows are split by ditches and quiet highways. Limestone cliffs, dun and mottled, loom over the landscape, which has long been an epicenter of agriculture. Around A.D. 450, while people along the southern river were still focused on fishing and foraging, villagers in the American Bottom began to systematically clear the local forests for the sake of what can only be considered farms.

For the next few centuries, the farmers here grew now-forgotten crops, including marshelder, knotweed, and maygrass, all of which, like sunflowers, grow starchy and nutritious seeds. By 900, these people added a new crop: corn. And as if drawn by this new species, people flocked down from the bluffs and into the rich riverlands. A single town in the American Bottom in this era could host a thousand residents. These people continued the ancient tradition of mound building, and the largest villages featured temples built atop flattened pyramids. The residents of outlying villages would gather a few times a year for feasts; councilors would come from various towns to work out regional issues.

The success of these towns and their corn farms drew immigrants, too—up from Missouri and Arkansas, down from Illinois, perhaps even from Caddo country, among other regions. These people were ethnically distinct, likely speaking different languages, though they shared certain spiritual customs. Some of the southerners may have been keepers of the ancient traditions that first inspired the mounds. And their coming together seemed to spark something new: Around 1050, someone decided to raze an old creekside village. In a matter of months, the houses were buried under dirt. Then came a wave of construction. Dozens of new neighborhoods appeared, thousands of homes, all built in standardized form. The walls were prefabricated; the roofs were hipped, woven from boughs, and reached nearly to the ground. Eventually more than 120 mounds were built within this new place, which can only be called a city—the first city in the watershed, indeed the first in North America north of Mexico.

Across the surrounding floodplain, fields of corn and squash and sunflower grew atop small ridges, tended by families of farmers. Their houses stood in long lines, one every few hundred yards. Weaving between the plots were old creeks and slough: this was still a wetland settlement, and sometimes the easiest way to travel between neighborhoods was by dugout canoe. The city center was protected by a wall of wooden palisades, two miles in perimeter. Inside these walls stood the new city's greatest architecture, the greatest display of engineering yet undertaken north of Mexico. Even the fifty-acre plaza—as big as forty football fields—was a marvel. The subtle topography carved by the river in earlier eras had to be leveled precisely to keep any water from pooling after a heavy rain. Just clearing this space would have taken a force of a thousand workers more than a month. The city's preeminent pyramid stood to the north of the plaza. It grew ever taller over the decades until it hit a hundred feet, the height of a ten-story office tower. The base spread over fourteen acres. The top was level and featured a large structure—perhaps a temple or a council house, or a residence for a chief. According to archaeologists' calculations, a speaker standing at the top could project his voice across the massive plaza.

Four sets of smaller mounds and plazas surrounded this central node, one in each cardinal direction, creating what archaeologists call a cosmogram. The entire city was carefully laid five degrees off the meridian—aligned with the site where, once every nineteen years, the moon rises at its southernmost point on the horizon.

Perhaps ten thousand people lived within the six-square-mile central city, with tens of thousands more scattered about other nearby villages and across the countryside. (A second, smaller precinct was built just to the west, closer to the river; another appeared across the Mississippi, where modern St. Louis stands.) In the United States, no city matched this size until the end of the eighteenth century. This place—now known as Cahokia—is one of the most important sites in the Americas, though its story remains in dispute.

~~~

Archaeologists uncovered one important piece of evidence when they excavated Mound 72, a rather inauspicious hump of land south of the central plaza. The dig revealed two wrapped corpses, placed amid a massive array of burial gifts. A two-inch layer of shell beads, for example, included more than twenty thousand beads in total, more than have been found across the rest of the city combined.

The greatest gifts were human. The shell beads surrounded two bodies, which lay buried alongside many other corpses, including fifty-three women carefully arranged in two rows. Another set of bodies, male and female, had their heads removed. Archaeologists concluded that these victims had been lined up along a pit and executed in sequence by being bludgeoned with a club. This burial seemed to echo ancient legends, passed down through generations of Indigenous people, and archaeologists began to weave a story: the two honored bodies belonged to warriors who ruled this city, men who had invoked the old legends and claimed to be the descendants of gods.

Despite the obvious importance of this city, Cahokia's boom did not last. After a fifty-year burst of construction and a peak of as many as fifteen thousand residents, the population halved over the next century. Some of the people who left the city settled into the surrounding farmlands, but after another hundred years, as the city's population nearly halved again, people began abandoning the region entirely. By 1300, just scattered families were left, and within another few decades the region from Cahokia south to Memphis—four hundred miles of river—would be mostly empty. ("The Vacant Quarter," archaeologists call it today.) Where all these migrants went is not clear. Perhaps some returned to ancestral homelands, up and down the river.

There are various hypotheses to explain the decline of Cahokia. Droughts, maybe, or perhaps the inverse. One theory suggests the Cahokians tore so many trees out of the American Bottom that the
~~~

resulting erosion launched devastating floods. But the soils show no record of such catastrophe. Corn has been blamed, too, on the presumption that the city so depended on the crop that a few bad seasons could spell its doom—though this, too, is contested by archaeologists who have found that Cahokia's feasts featured a diverse array of crops. Archaeologists increasingly view ecocide, as you might call such stories of human-induced catastrophe, as a Western projection, a false presumption that human cultures are inherently destructive to the natural world.

It's likely that the region was politically unstable as Cahokia declined. Walls were coming up around towns and villages, though Cahokia's central districts don't seem ever to have been raided. It's possible that this place had lived its natural lifespan as a chief-based nation. Anyway, the end of Cahokia did not mean the end of the customs that emerged here. The site is seen as the birthplace of what archaeologists call the Mississippian culture. It's marked by fortified cities built around mound centers, by striking visual iconography—fork-tongued serpents, a "birdman" warrior who is part human and part raptor—and by the presence of a class of elites who inherit their status. As Cahokia waned, this way of living spread into new floodplain valleys—up the Tennessee and Cumberland rivers, for example, into Kentucky and Tennessee. When Hernando de Soto arrived in 1539, he found Mississippian-style cities as far east as Georgia and South Carolina, beyond the watershed. (Along the Lower Mississippi, meanwhile, where mound building began, his men found Quigualtam and his troops. They were a part of the Plaquemine culture—people who, as we'll see, despite sharing certain traits with the Mississippians, were notably not Mississippian.)

The first time I visited Cahokia, I decided to watch the sunrise from atop the central mound. It's known as Monks Mound, thanks to a Trappist monastery that was built nearby in the 1800s. (On Monks Mound itself, the Trappists grew produce on soils they knew would never flood.) In the 1940s an entire subdivision appeared around the Cahokian earthworks, sixty suburban houses, including, eventually, a backyard pool installed at the foot of the great mound. Today the homes are gone; just scattered trees remain, marking the edges of the former streets. The central plaza and its surrounding mounds are preserved in an Illinois state historic park, but the region beyond offers a compendium of sad Americana: warehouses and trucking depots, steel refineries and coal-loading terminals, everything abutting the channelized remnants of ancient creeks. The check-in kiosk at the Indian Mound Motel is protected in glass; before it went out of business entirely, the Mounds Drive-In had switched over to X-rated films.

As I waited for the first rays of light, a trio of locals appeared and told me they had just wrapped up a long night wandering from bar to bar. Looking to the west, I could see another mound—an actual landfill. It opened in 1950, and five decades later, it was as tall as Monks Mound. Before I'd come, friends had told me that to watch the sunrise here was to experience a deeply spiritual moment. But in this setting, with this company, I did not feel much of anything. Only later did it occur to me that, by climbing to its top, I might be misinterpreting this place. The point of the monument may not have been to serve as a place where one could stand, alone on a precipice, looking down on the world.

The insights of archaeology are at once enticing and frustrating. We can know the material facts of the continent's ancient history with startling clarity. The teeth of long-dead Cahokians reveal the precise generation when people began eating larger quantities of corn. Wooden residue, preserved in the soil, shows us the shape of the city's long-lost neighborhoods. Fossilized pollen tells us when the surrounding forests were chopped down. We can get a sketch of this place, yet the picture remains so empty. The questions I most want answered cannot be easily calculated from material remains. What did these people call themselves? What stories did they tell as they poured out their basket loads of soil to build their mounds? What did they know about this river that we, too, might want to learn?

There is another way to study the mounds, as Chadwick Allen points out in *Earthworks Rising*. We can see them as a form of writing—as

"Indigenous knowledge encoded in the land." Indigenous stories of earthworks often emphasize what Allen calls the "three-worlds theory." There is the Upper World, above us, which features order, or at least a cyclical permanence: the sun and moon rise and set, an endless dance precisely repeated; the stars, too, for all their seeming boundlessness, rotate in a complex but predictable swirl. Below us is the Lower World—a chaotic domain of water, ever changing, but also abundant and fertile. It might be tempting to equate the Upper and Lower worlds with the familiar Western concepts of Heaven and Hell, but neither is preferable and both are necessary. This worldview upends our modern notions of disaster: A flood is not a catastrophe but an asset. Fish can pour out of the river, into the reconnected backchannels; fertile new soils are dumped across the land.

The Choctaw novelist LeAnne Howe has noted that these realms can overlap, forming spiritual ecotones. The sites where mounds appear are often ecotones, too: the earthworks stand upon the edge of wetlands, as if perched alongside the Lower World; they rise toward the sky, aligned precisely to match the turns of the Upper World.

We live in the third part of this cosmos, the Middle World, an island in between. Some Southeastern people tell what are known as "earth diver" stories, about how this world came to be. A creature—a muskrat in some stories, a crawfish in others—swims down into the endless waters to bring up a handful of mud. Howe describes how early mound building might have been "a kind of theatrical performance, one in which the performers tell the story by collectively sculpting the earth." The mounds, then, can be seen as a renewal, a reenactment of the world's beginning.

If you rotate this cosmological model and lay it across the Southeast, the landscape embodies its attributes. (It's a "geographic onomatopoeia," as one archaeologist has put it.) To the north, mountains and plateaus—the Ozarks, the southern edge of Appalachians, the Piedmont—reach into the sky. To the south, the land drops away into a tempestuous, hurricane-churned gulf. The Mississippi becomes a kind of cosmic river, as the archaeologist Kenneth Sassaman puts it, a link between these Upper and Lower worlds.* He believes that Poverty Point, the first great monumental gathering place on the continent, may have been a portal into the underworld.

This idea is not universally accepted; Sassaman jokes that his colleagues must think he's dropping acid. Mounds are sometimes analyzed for their biological implications, turning the science into a kind of evolutionary ecology: *How many calories would be burned in building a mound? What kind adaptive advantage might this construction provide?* The point is to depersonalize the earthworks, to stick to what can be quantified. But Sassaman prefers to engage in a different kind of study. He imagines the people who dwelled along this river *as people,* hypothesizes about their thoughts and beliefs and intentions. Then he seeks ways to check his conclusions against the facts on the ground.

Here are some facts, then: the process of construction at Poverty Point, which lasted a few centuries, began around 3,400 years ago, when a set of concentric half-circles were built. These enclosed the plaza that opens to the east—toward the riches of the river, and toward the sun that rises each day out of the Lower World. Later, a series of mounds were added in a line to the west, including the site's largest mound, which, according to many observers, looks something like a bird flying toward the setting sun. Beneath its base is an old wetland, a bit of the fecund Lower World. The builders appeared to have never paused and never been interrupted by a storm. That means the work proceeded incredibly quickly; in the historical era, the longest this region has gone without heavy rainfall is thirty days.

The site these people chose seems to be an intentional nod to the past. A mile to the south of the largest mound, there is a smaller, older

* The Mississippi is not the only cosmic river. Sassaman suggests it might be one link in a wider network that mimics the circular flow of the sun. After the Mississippi empties into the Gulf of Mexico, the current flows east toward Florida—a region that has been linked by artifacts to the prehistoric Mississippi River. There the waters of the St. John's River percolate upward through limestone, then flow north into the Atlantic. Finally, a bit farther north in the Carolinas, one more river completes the loop: the Tennessee pours west, eventually rejoining the Mississippi.

earthwork that had been abandoned a thousand year earlier; a line drawn through Poverty Point's biggest mounds runs precisely through that predecessor. This site, then, seems to be a map of many dimensions—an atlas of both space and time, referencing the people and places and forces that preceded its construction.

There were scattered additions after the largest mound was finished. Large posts were planted in the plaza, marking out circles; then a final, smaller mound was built just to the north. At some point, hundreds of pieces of broken soapstone vessels were buried; the materials had been carried from the Appalachian foothills in Alabama or Georgia, which almost certainly meant an eight-hundred-mile journey by water, including a long passage at the edge of the Gulf of Mexico. Soapstone burials were an element of the era's funerary traditions—but here, alongside the largest known cache of such materials, no bodies have been found. That absence seems key to me. So too does the fact that the last rounds of this work occurred amid an era of climate-driven crisis. "Rapid and catastrophic flooding would have rendered much of the alluvial valley uninhabitable for prolonged periods," the archaeologist T. R. Kidder notes.

Within a century of the completion of the last mound, activity seemed to cease at Poverty Point entirely. If anyone still arrived on pilgrimage, they left behind nothing. Life seemed to change everywhere along the southern river, in fact. Traders abandoned their long-distance routes. After centuries of gathering, people settled into quieter, local lives. Mound building continued, though on a far more modest scale: mounds became the focal point of a village rather than a continent. So if there was a funeral here, perhaps it was not for a single person, or even for a people, but for a way of living no longer feasible on this earth. Perhaps the entire, centuries-long building process was a way of saying goodbye.

~~~

The nation's founders were obsessed with the earthworks. Thomas Jefferson dug bones out of a burial mound near his Virginia property. Early settlers along the Ohio River built their towns at the foot of ancient monuments, as if to signal that the new villages, too, were destined to be grand.
~~~

But since the settlers' claims to this land depended on the idea that the local people were wasting its potential, they had to concoct wild theories to explain the presence of these old monuments. Some supposed that Soto and his men might have built them. Others suggested that Vikings were the architects, or wandering Israelites, or Welshmen, or Egyptians. What was clear to the settlers, at least, is that whatever advanced thinkers had built these structures were gone, wiped out by the barbarous peoples who now occupied the continent.

It was not until the end of the nineteenth century that western scholars put this mound builder myth to rest and acknowledged that the ancestors of Native Americans had indeed built the earthworks. Even decades later, short-sighted assumptions persisted. When archaeologists first noticed the concentric ridges at Poverty Point in the 1950s—revealed thanks to aerial photographs—most presumed such extensive earthworks had to have been built after the advent of farming. How else could people have stored the food required to build such extravagances? They were shocked when its true age was revealed. They were shocked again in the 1990s, when they learned that the mounds at Watson Brake were another two thousand years older still.

It turns out that the swamps and oxbows were just that lively; massive corn farms were not needed in order to build. The Mississippi River, then, requires us to rethink the whole idea of civilization: what it means and how it began. In Western culture, we sometimes think of time as a one-way line. Societies walk up a staircase, settling into some chosen homeland where people learn to garden and eventually farm. They build villages and then towns and then cities. In North America, though, the first great cultures did not begin by taming the wilderness. They flourished *because* of the wilderness in these swamps.* Millennia later, the swamps' abundance still set the place apart. As corn farming swept across

* This is not just the case in North America. In the Middle East and Africa, some of the first sedentary communities were settled by nonagricultural people and situated, just like the Mississippi mound sites, on the edge of wetlands. These were ecotones, border zones, where hunting-and-gathering people could access the resources of multiple ecosystems and therefore not worry if something—a flood, say—caused problems in one of them.

the American Bottom, the people living below the mouth of Arkansas River, just a week's paddle downstream, took little interest in the crop.

By the time Cahokia waned, the people along the lower river were planting corn. Still, the place seems different, separate. Local graves did not feature lavish burial gifts, suggesting a rejection of the kind of social hierarchies that marked Cahokia. There was trade between the Mississippians and the southern people, known today as Plaquemine, but in some regions, especially in Louisiana, people "ignored or maybe even actively resisted the lifestyles or worldviews of Mississippian culture," archaeologists say. They looked instead to the ways of their own ancestors, who had been building mounds—and fishing in these swamps—for thousands of years.

Despite this resistance, the southern river at some point became associated with agriculture: across the Southeast, and as far away as North Dakota, there are stories of Grandmother, a spiritual figure who taught women how to raise the continent's first crops. In some versions of the story, she is linked to the Lower World, and with the Mississippi River in particular; by bathing in its waters she could renew her youth again and again. On her delta island home, Grandmother is protected by a water snake—from whose body, in fact, she coaxed sunflowers and squash. For centuries, farmers sought her guidance and protection so they'd have good weather and bountiful harvests. It seems that this woman, or at least a similar figure, known to archaeologists as Earth Mother, was revered in Cahokia. Figurines depicting her and her serpents have been found at nearby sacred sites. The richness of the southern swamps, the place where so much began, became woven into cultures across the watershed.

Strangely, Indigenous oral histories contain few references to anything that might resemble Cahokia, even among the people seen as the Cahokians' likeliest descendants. The closest may be Osage tales that describe their migrations: one version suggests that as their forebears wandered the Middle World, they happened upon a people who destroyed whatever life they encountered, and so lived in a village surrounded by bleached-dry bones. By joining the Osage on a westward migration, these people—the Isolated Earth clan—were able to find a new home free of so much death. Cherokee stories, meanwhile, tell of an ancient caste of priests who grew so powerful that even the Earth rejected them, spitting out their bodies. Burial mounds offered a way to contain their problematic remains. Such stories suggest that people may have looked back at Cahokia and its violence with dismay and regret.

Still, the idea that Cahokia was a brutal place is now contested. There are questions about how centralized its economy ever was: whoever ruled this place was supposed to have sustained their power by doling out surplus crops, but no one has found the central granaries this would have required. Sure, there were elites and commoners; some people had better food and more possessions. But farming "wasn't controlled by the muckety muck chiefs," archaeologist Gayle Fritz told me when she toured me through the site. Fritz, who focuses on ancient agriculture, has studied the role of sacred female figures like Earth Mother. She noted to me that a decade ago, the two honored bodies in Mound 72, long assumed to be men, were revealed to be more likely a man and a woman. Hidatsa and Mandan women, whose ancestors likely settled along the Missouri after heading west from Cahokia, tended crops; if they were good at it, they could join a society that made key agricultural decisions. Perhaps if Cahokia really was a brutal place, this custom developed as a reaction to its authoritarianism—a new political structure meant to prevent future abuses. Or perhaps it's a tradition that began at Cahokia itself, where, given the importance of Earth Mother, women likely sat in positions of power.

Certainly the ritual enacted at Mound 72—the dispatched bodies, the celebration of the dead—was meant to tell a story. But the story may have been about unity rather than power. Some archaeologists have always believed that the point of the burials here was not to aggrandize specific individuals but to honor ancestors at large, who in death all become equal. The massive feasts held at Cahokia, sometimes assumed to have been elite-sponsored events, might have been communally hosted, with each family contributing food as a gift. As for the mangled corpses, sacrifice can be cruel, but it can also be honorable. Think of the young men and women who volunteer to join the U.S. military. In the Cahokian cosmos, to submit to a sacrificial ritual may have been a path to achieve distinction, to give the greatest gift to the rest of the beings in this dangerous island of a world. The anthropologist Jay Miller spent years studying mound-building traditions, including modern busk ceremonies in Oklahoma, where the descendants of many of the riverside tribes now live. They still build mounds, albeit on a smaller scale: at ceremonies, congregants stomp and sing and renew their world by assembling a fresh pile of soil. Miller concludes that practices that academics interpret as "tools of political competition and control," are, in Native teachings, more often an effort "to survive in a very unsteady world by means of community-based action done in prayerful manner."

It's too easy to think of the ancient earthworks as relicts, ruins that offer a glimpse of a former continent, a lost world. This is wrong: Indigenous people carry dirt from reservations in Oklahoma to add to their "mother mounds" in the Southeast. Construction continues. Besides, in Indigenous thought, progress is not a one-way line. Grandmother bathed in the Mississippi, becoming young again; so too did people build and rebuild earthworks before giving them up, turning to quieter lives, and then building again. Thousands of years passed, sometimes, between the most monumental of constructions. There is no lost world. The earthworks are patient things. Their timescale lasts centuries, and their story is not over. It's just that for a few centuries too few people have been listening.

Part II

AMERICAN RIVER

1803 TO TODAY

The monument at the Initial Point (Monroe County, Arkansas).

4

Half Horse, Half Alligator

The river in its not-quite-American years

No bright line marks the beginning of the river's American era. But one good place to start this part of the story is October 27, 1815, when Prospect Robbins moored his boat at the point in the alluvial swamps where the Arkansas and Mississippi rivers meet. He set a post to mark his arrival, and then, along with two assistants, he began to trek north into the muck. On the same day, a second man, Joseph Brown, embarked with a separate team at another confluence: the mouth of the St. Francis River, fifty miles to the northwest. Brown headed west. Both groups were official emissaries of the U.S. government, there not just to scout the landscape, like the earlier explorers, but to lay upon it a perfect rectilinear grid.

Both of these surveyors were veterans—Brown a captain, Robbins a lieutenant—of an army that just a few months earlier had been fending off a British invasion. Now, with the war over, their former commander had picked these two to complete the next essential task. A massive effort to survey the federal lands had begun two decades earlier; it was ready now to spread into the western watershed. Robbins and Brown were instructed to establish the "initial point" for the era to come—a reference site around which all subsequent surveys across the territories could be arranged. Think of the origin on the coordinate plane in a graph in an algebra book: a zero-point, a center place, arbitrary but precise. From this beginning, the great wilderness would be tamed into its map.

Brown and Robbins assembled teams of men they knew and trusted to serve as chainmen and axmen and markers. They were instructed by their superiors to leave late in the year so as to avoid the "inundations, the undergrowth, weeds, & Flies of various descriptions." ("No mortal man could take the woods before October," one official added.) They slept in tents and lugged drinking water in pails. These frontiersmen—former salt traders, in a few cases—knew how to hunt in the forests, what gear to carry, how to create quick shelter. But don't picture grizzled mountaineers in stinking buffalo robes. These men were well educated—the polite, church-going neighbors down the block. Robbins was a former schoolteacher. It's just that in this era, in this place, you needed to be hardy to make something of yourself.

As they trudged through the Arkansas swamps, they lugged chains to measure their progress, marking their passage at half-mile intervals, typically by driving stakes into the muddy ground. At the end of each mile, they selected some stout trunk and carved a mark to make a "witness tree." And on November 10, after two weeks in the wet forests—plodding forward just four miles on average each day—Robbins must have spied some sign of Brown's passage. He slashed two trees, indicating the point where the two lines crossed. The initial point was set. Robbins wrapped up his assignment two months later, hundreds of miles north, on the banks of the Missouri River in the Ozark hills. Brown traveled on across the floodplain to reach the Arkansas River. But in all thirteen states that would emerge from the Louisiana Purchase, as far away as Montana and Minnesota, parcels would be oriented around these trees.

Not that the initial point had much to recommend itself. Brown described the terrain around the site as "low," featuring "cypress and briers and thickets in abundance." He seemed unimpressed, repeatedly grading this floodplain territory as second-rate land. Robbins, too, had his doubts: when his former general offered a patch of the Arkansas floodplain as a part of his "war bounty," the surveyor declined, figuring it would be too much work to wring out a profit.

In the 1920s, a new group of surveyors arrived, in an attempt to clarify the local county boundaries. As they hacked through the overgrowth, they noticed Robbins's slashed trees and realized what they'd found. The locals decided to preserve this place, so today it remains a tiny island of swamp amid the surrounding sea of soybeans, a little-visited state park. A boardwalk allows the few tourists who do come to navigate across the wet soils, to the place where, according to an official placard, "the settlement of the American West began."

Prospect Robbins, born in Massachusetts, might have balked at the idea that he was only just reaching the West. Even Pittsburgh, the jumping-off point for many journeyers headed downstream, was considered part of the "western waters." There in Pittsburgh, in 1801, the savvy printer and bookbinder Zadok Cramer had published the first edition of his great success, *The Navigator*—a mile-by-mile guide to the Ohio and Mississippi rivers. Updated every few years with data gleaned from letters sent east by settlers, Cramer's book provides a portrait of the watershed in these first, not-quite-fully-American years. "This noble and celebrated stream," Cramer writes of the Mississippi, "this Nile of North America, commands the wonder of the old world, while it attracts the admiration of the new."

A contemporary river traveler, accustomed to a deep and wide ribbon of water, may find it hard to envision the cluttered waterways that Cramer describes. The debris began at the Mississippi's mouth, where shoals blocked the three forking passes that lead out into the Gulf of Mexico. Colonial pilots would sometimes press their boats into the mud, then unload their cargo while awaiting a tow. Similar obstacles marred every tributary. Sandbars and rock bars and gravel bars could be broken down into a full taxonomy describing their size and shape and orientation: chains and traps, riffles and reefs. The largest hazards earned names all their own, which combine to make a rough American poetry. Big Bone, Pig's Eye. Glass House, Scuffletown.

Where the rivers wound through softer soils, the banks crumbled easily. Whole trees—ancient, massive sentinels that might weigh as much as sixty tons—shed into the water, sometimes hundreds at a time. The sound, according to a later federal report, resembled "the distant roar of artillery." Once in the channel, the roots grew matted with dirt and cobblestones and implanted in the river-bottom mud. The resulting hazards were known as snags, and they, too, inspired a full lexicon. "Planters" sat immobile; "sawyers" bobbed in the current; "sleepers" lay wholly beneath the water. In places, the driftwood had gathered into a thick masses as big as islands. The effects of all this wood could be deadly; as many as a quarter of the flatboats wrecked en route to New Orleans. Among Cramer's "instructions and precautions," he emphasized the importance of selecting a quality vessel. He recommended a certain vessel in particular: a large wooden raft, typically somewhere around sixty feet by fifteen, with a wooden box on top that served as makeshift quarters. The raft was known by many names—*ark*, *broadhorn*—but the term that stuck was *Kentucky flatboat*, in honor of the place where so many trips began.

Some dangers could not be solved by picking the right boat: "counterfeiters, horse thieves, robbers, murderers, &c," as Cramer put it. So many stories were spun that it's hard to distinguish truth from fiction now. Cave-in-Rock, a riverside cavern near the mouth of the Ohio River, became a particular focus of blood-soaked tales. Here crews of hardened criminals were supposed to have enticed travelers with decoys—attractive female compatriots who asked for a ride south, or a sign that advertised "Wilson's Liquor Vault and House for Entertainment." The victims were said to be murdered, their cargo hauled downstream for sale by the pirates. Indigenous soldiers were also a worry: they viewed the flatboats as part of an imperial invasion, enemies they needed to stop if they wanted to hold their homelands.

Indeed, Thomas Jefferson had, if not quite a plan, then at least a vision for this great uncharted landscape: the Mississippi Valley would fuel the creation of an "empire for liberty." He first used a version of this phrase in a letter to George Rogers Clark during the Revolutionary War, in which he urged the former surveyor to head north to wrest more land from the British. The frontiersman proved unable to muster sufficient recruits, but Jefferson never dropped the idea.

It's worth pausing a moment to describe Jefferson's dream, because even if the phrase itself—*empire for liberty*—has not survived in our modern vocabulary, this is a Mississippi River idea that haunts our national consciousness. Jefferson turned the land hunger that drove the first western settlers into a political imperative: democracy could truly flourish, he claimed, only if the new nation expanded across a substantial territory, ensuring that as many citizens as possible could own a portion of the land. Jefferson believed men who lived and worked on their own property could never be trapped in a dead-end job in one of the new textile mills popping up in Atlantic cities. They'd remain unyoked and unencumbered, with no debt and no boss. They would become perfect democratic citizens, free to decide for themselves. And the Mississippi watershed offered the space that Jefferson needed to establish an empire of small, landed farmers—"cultivators," as he called them, or "husbandmen."

This all doubled as a vision of the ideal landscape. Jefferson did not want a collection of the soot-stained, over-mobbed cities that were growing like "sores" on the body of Europe. Nor was untamed nature a suitable fit for the new nation. He hoped the watershed would be converted to a garden—or a collection of gardens, spreading across the landscape like a quilt. Private property everywhere. The only shared resource he spoke of was the river itself, the highway into his promised land.

Jefferson saw it as a "law of nature" that anyone who lived along the banks of a river ought to be allowed to travel its length. The first settlers in the Ohio forests seemed to agree, as the flatboat rush commenced decades before Louisiana changed hands. This was, after all, a far easier voyage than lugging crops over the mountains to Philadelphia or New York. By the early nineteenth century, in the wake of the Louisiana Purchase, as many as three thousand flatboats traveled down the Mississippi annually, carrying the goods of a young nation: pine planks, pork, flour, whiskey and tobacco, hemp and rope and sacks, cattle and horses, cotton, peltry and lead, cutlery, ears of corn, barrels of apples and potatoes and cider and dried fruit.

The boats drifted atop an ever-changing river. The Upper Ohio was transparent, revealing boulders below in its channel. Then after a few hundred miles, the terrain flattened; mud thickened the water, until it was a torrent of half-milked coffee. Even the fish in these waters seemed ungodly: catfish could weigh a hundred pounds. On some nights, they slammed against the boats so loudly that it was hard to sleep.

When travelers interested in acquiring furs reached the Mississippi River, they might head north, past the string of old French villages founded by priests living among the Inoca, to reach St. Louis. Established in 1763, the former trading post had grown into a frontier crossroads, two hundred homes perched atop the bluff where the Missouri and the Mississippi meet. But the bulk of the traffic headed south, spending weeks winding through a valley that was flat and wet and mostly empty. Besides Natchez, no substantial settlement appeared for nearly seven hundred miles. Finally, in the delta, the levee appeared, protecting the plantations. The final seventy miles into New Orleans featured orange groves and sugarcane fields and mansions, like one long and lovely village.

Those who made it safely sold their wares in New Orleans, then sold the warped wood from their flatboats as scrap. (Often the planks were laid atop the city mud to make a sidewalk.) The crews walked home overland, following a set of Indigenous trails known as the Natchez Trace. They traveled in packs of twenty or more to avoid being robbed. If a farmer bought anything bulky with his profits, it would have to be sent north by keelboat—a long and narrow vessel with a pointed bow and stern, which at the time was the only way to carry any substantial cargo against the current. Typically sixty feet long and eight feet wide, capable of bearing forty tons, the keelboat was specially designed for the western rivers. Still, an upstream trip would require the muscle of at least ten men.

If a keelboat crew was lucky, they could unfurl the sails to exploit a favorable wind. Otherwise, the work was wearying. Sometimes the boat's

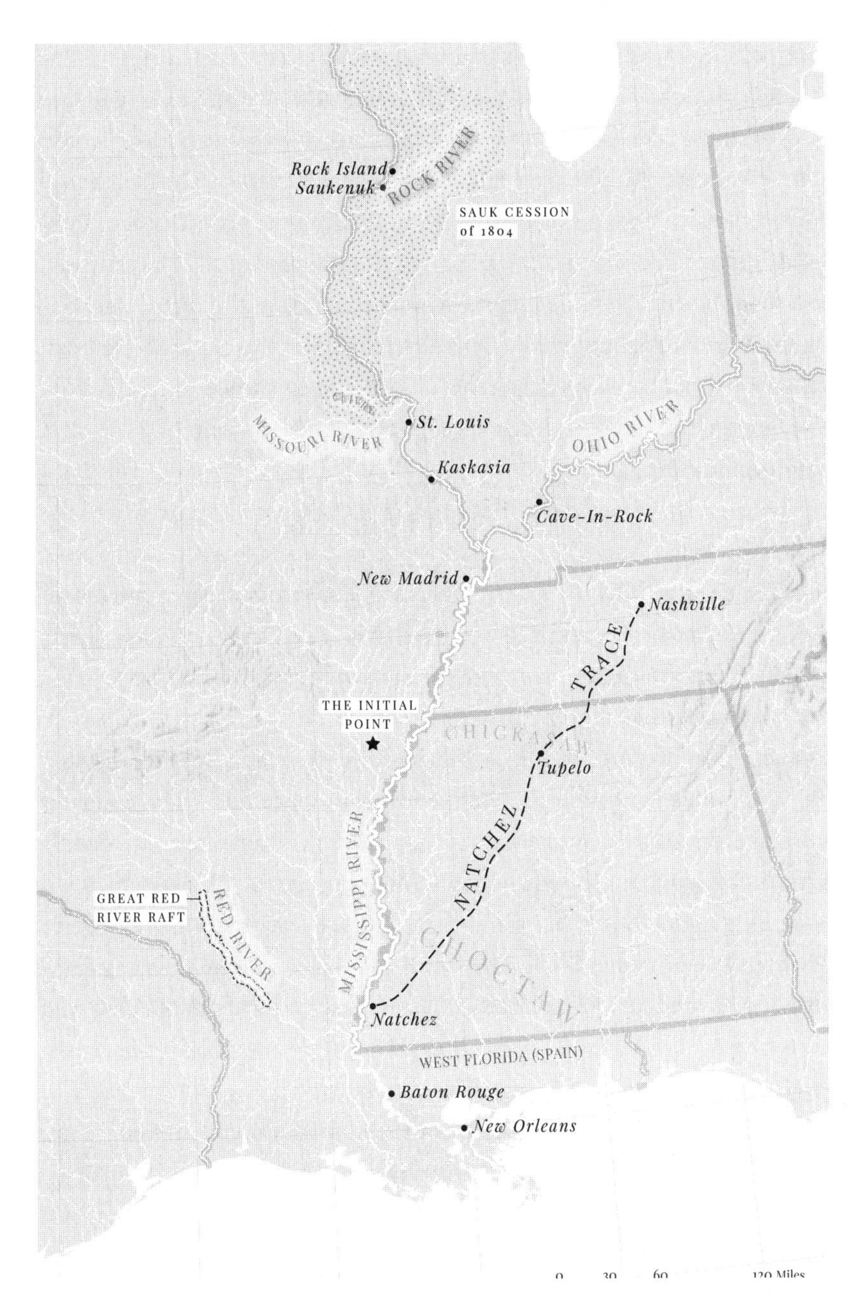

The western rivers in their frontier years.

best swimmer would head to shore with a rope clamped in his teeth. The rope was attached to the mast, and the swimmer tied the loose end to a tree; then the crew dragged the boat forward, one thousand-foot length at a time. During floods and high water, keelboat crews grabbed at brush and branches along the shore so they could drag the boat forward. Typically, though, the men jammed spears into the mud at the riverbottom, and then, bracing their shoulders against a crutch at the top of the pole, walked forward on the narrow planks that lined each side of the boat. When a boatman reached the front of the line, he pulled his spear free, then hopped atop the cargo box at the boat's center to sprint to the back of the line and start again.

The keelboats hugged the inner bends of the river's curves, where the water was slower, though this meant an arduous crossing after each bend ended and the next began. Often, a keelboat could manage just two crossings a day, for a total of fifteen or twenty miles; afterward, their shirts bloodied, their shoulders callused, the men were rewarded with a *fillee*—a cup of whiskey chased by a cup of river water.

These men lived a life that was, according to one traveling preacher, "in turn extremely indolent, and extremely laborious." The indolent moments sound pleasant enough. When the boats were moored, a fiddle was always playing. The music led to dancing and drinking—which led to cussing and fighting, in legendarily elaborate fashion. On a voyage in 1808, a traveler named Christian Schultz descended to the squalid neighborhood at the foot of the Natchez bluffs, known for its flophouses and gambling dens, and found himself captivated by a handful of boatmen caught in a "dispute respecting a Choctaw lady."

"I am a man; I am a horse," one of the drunken men hollered. "I am team. I can whip any man *in all Kentucky,* by G-d."

The other upped the ante: "I am an alligator," he said. "Half man, half horse; can whip any *on the Mississippi* by G-d."

The men "went at it like two bulls," Schultz wrote, "and continued for half an hour, when the alligator was fairly vanquished by the horse."

The image of the boasting boatman became a literary trope, and one boatman emerged as a particular source of fascination. Mike Fink was a real man, but the stories about him are clearly exaggerated. He's made to sound like a hunk in a romance novel—heavily muscled, symmetrically proportioned, so frequently shirtless that his skin had darkened, and he was sometimes mistaken for an Indigenous warrior. He was a crack shot and a "helliferocious fellow," as one story put it, "and there aint a boatman on the river, to this day, but what strives to imitate him." An influential account of Fink was published in 1828 by Morgan Neville, who is now remembered as "the first notable writer of fiction to be born west of the Alleghenies." Neville, a Pittsburgher himself, was likely honest in his claims that he crossed paths with the legendary frontiersman.

Neville suggests that Fink got his start as a scout in the Upper Ohio River watershed, living "as did the Indian," spending weeks alone in the woods, eating parched corn instead of bread. He slept under the stars, rolled in a blanket. Such scouts served as the advance forces for white conquest, monitoring Indigenous soldiers, ready to warn nearby settlements of any hostile approach. But the scouts themselves were often the aggressors. In a telling anecdote, Neville suggests Fink shot an Indigenous hunter for the simple offense of stalking a buck that Fink hoped to kill. In 1795, after ninety-nine chiefs signed a treaty that gave Ohio to white settlers, the scouts were out of business. By then, apparently, Fink's lifestyle had left him unsuited to a settled home, so he committed to a life on the river. For Neville, Fink served as a fitting emblem for an era—an aspirational idea for thousands of young men, some just farm boys chafing under the glare of their fathers, feeling drawn to the motion of the river.

Jefferson was hoping for a more decorous sort of culture along the river, though his own dream contained plenty of violence. It's often overlooked that what Jefferson purchased in Louisiana was not the land itself, which the French had not yet fully acquired, but rather the right to negotiate *for* the land. With the exception of a few tracts recently acquired from the Choctaw and the Kaskaskia, the United States could not claim even the territory along the Mississippi's east bank. In a letter to William Henry

Harrison on the eve of the purchase, Jefferson laid out his preferred strategy for getting it all.

Americans should encircle tribal villages with settlement, he said, choking off their hunting lands and thereby forcing a complete dependence on agriculture. Then the government could establish trading posts and "be glad to see the good & influential individuals among them run in debt." These measures would drive the Indigenous people to sell some land, Jefferson figured. And should anyone object and "take up the hatchet," as Jefferson put it, the president was clear: they should be crushed. In 1804, just months after Louisiana changed hands, a group of Sauk hunters took up the hatchet: they killed three settlers along the Cuivre River, in the Ozark foothills, northwest of St. Louis.

A French priest, upon encountering the Sauk more than a century earlier, had concluded that their warlike nature meant that "above all others [they] can be called Savages." The most famous Sauk, Black Hawk, first went to battle as a teenager, as his forefathers had before him. Family mattered to Black Hawk: he could trace his lineage back generations to a great chief who warned his people of the coming troubles with the settlers. When Black Hawk first watched his father, Pyesa, kill an Osage enemy, the sight lit a fire in the young man. Black Hawk raised his lance and quickly made his own kill, then cut loose the dead man's scalp. Pyesa looked on in silence. But Black Hawk thought he could see in his father's countenance some sign of pleasure. For decades, even after Pyesa's death, Black Hawk strained to prove himself a worthy and courageous son.

After that first battle, Black Hawk became an admired soldier, a leader capable of rousing armies, though this was not his whole self. A few years later Pyesa fell in battle against the Cherokee; Black Hawk blacked his face, a sign of his mourning. He took a five-year pause from warfare. This entire nation had this other side: the Sauk were not just warriors but productive lead miners and talented farmers. When a late eighteenth-century explorer stumbled into their largest village, he remarked on its tidiness. Ninety lodges built from wood, lined in bark, were laid out around a gridded arrangement of wide streets—"more like a civilized town than the abode of savages."

By Jefferson's days, the Sauk had moved on to a new village, built near the mouth of the Rock River. This was a fine place: lush bluegrass riverbottoms offered pasture for horses; rapids in the Mississippi trilled with fish; prairies along the big river served as farm fields. Saukenuk, as they called the village, was a place to gather, dance, and celebrate, to raise crops and feast on their bounty. Through the warmer months, the Sauk lived in the longhouses; each fall they dispersed into their hunting lands. That was why, in September 1804, four Sauk hunters crossed the Mississippi River.

The Sauk, Black Hawk included, were wary of the Americans. The stories they'd heard about this new breed of settlers were not encouraging. So when the hunters found white settlers on the Cuivre, on land not yet ceded to the United States, they killed the invaders. The deaths put the entire frontier on edge.

The Sauk chiefs attempted to defuse the tension, condemning the attack, offering gifts in recompense. A few chiefs attended a conference with William Henry Harrison, the territory's governor. There are no records of what transpired at the meeting, but the United States emerged with a new claim to 51 million acres of land in Illinois, Missouri, and Wisconsin. Some was not the Sauks' to sell. One good hypothesis is that the Sauk thought the treaty was a symbolic gesture, an acknowledgment that now the United States, and not Britain, would be the imperial presence looming over their lives. The text indicated that the tribe would be permitted to hunt on the land for as long as it belonged to the United States. Perhaps the Sauk did not yet realize that the United States did not plan to keep the land. Instead, they'd sell it to private citizens to build Jefferson's empire.

The French approach in the delta had been to dole out "long lots" to its colonists, thin ribbons of land, each with its own small bit of river frontage. This was a sensible system in a region where the only mode of travel was the water. But an empire of liberty demanded fuller coverage. By 1785, the nation had settled on a plan, largely developed by Jefferson, to split the nation into his dreamscape: lay out a perfect grid of townships, each covering thirty-six square miles, which could be broken into 640-acre sections, then split again into 160-acre quarter sections. Rather than a free-for-all where an unorganized tumble of men like George Rogers Clark could pick lots on their whim, the whole of the empire would be a cataloged grid, its varied pieces sold at auctions in land offices established across the territories.

The task wasn't easy. When the secretary of the treasury asked in 1811 where in Concordia Parish, Louisiana, a land office might be established, a local surveyor counseled against establishing an office at all, at least for that year. A bad flood had liquefied forty miles of land, and even after the waters receded, there would be hazards to contend with: "the poisonous effects of Half dried mud, putrid fish, & Vegetable matter—almost impenetrable cane brakes, and swarms of mosketoes," the surveyor wrote. Floods walloped the region again each of the next two years. For the time being, then, the alluvial valley remained the quiet landscape it had been since the fall of the Mississippian chiefs.

To the south, there were rival claims to deal with. Spanish Florida officially extended west to the banks of the Mississippi. Congress secretly funded a group of rebels, who in 1810 attacked the handful of ill-prepared old Spanish soldiers who were occupying the fort in Baton Rouge. The rebels claimed the city for the United States. The claims of the many French and Spanish homesteaders who lived scattered across the backcountry were not so easy to extinguish. In the new state of Louisiana, so many old allotments cluttered the maps that the first public sales could not be held until 1820, after the government sorted the valid owners out from the squatters.

The specter of insurrection and war, meanwhile, made the nation's grip on the land feel loose. At the top of mind for many new settlers in Louisiana was the haunting story of how the watershed had turned American. Napoleon had planned to grow food along the Mississippi, which would feed the workers on his Haitian sugar plantations, whose crop would be shipped to Europe. Then the enslaved Haitians revolted, successfully. Without the island, the Louisiana farms were moot. Now, with the land in U.S. hands, more and more enslaved laborers were arriving along the river—raising worries that the same fate might befall the American claim. When in early 1811 an army of Black and Creole men arose along the Mississippi River, thirty miles upstream of New Orleans, this was a nightmare coming true.

Armed with machetes and pitchforks, they seized a cache of muskets, then burned down a mansion. They had likely gathered in the swamps behind the plantations to plan this attack; they must have known that the government was distracted by the ongoing fight over Spanish Florida. The rebels waved banners and marched to a drumbeat, sacking plantations as they descended on the city, recruiting more soldiers at every stop. In their wake, they created a zone, thirty miles long, where emancipation became, if not the law, then the fact on the ground.

The response was swift and strong: within a few days, the U.S. Army, working in concert with a local militia, routed the uprising. One participant called it "une grande carnage." The decapitated heads of some of the rebels were placed atop pikes along River Road, a reminder, to anyone else contemplating freedom, about who was in charge.

To the northeast, the increasingly dense settlements along the Ohio River had their own worry: a looming war against Britain. But in a sense the war was already underway. Many Indigenous people, subscribing to the theory that the enemy of my enemy is my friend, had allied with the crown. Two Shawnee brothers—one a prophet, the other a soldier—set up the headquarters for a burgeoning anti-American movement in the unconquered territory along the Wabash River. Late in 1811 a frontier militia led by William Henry Harrison burned the village. The battle is sometimes called the unofficial start of the War of 1812.

The next year, when Congress made this a proper war, Black Hawk knew which side to join. Soon, along with a battalion of Sauk and Winnebago soldiers, he took part in an attack on a small U.S. fort just downstream of Saukenuk. Two years later, when the U.S. Army sent a fleet upstream with plans to demolish the great Sauk village, a group of a thousand Indigenous fighters drove back the boats. To the west, traders were reporting that the rivers throughout the Missouri Valley were "shut against" the Americans, too. Despite the Louisiana Purchase, then, the watershed could hardly be called U.S. land.

The British, too, had been successful, stymying every U.S. attempt to invade Canada, storming Washington, burning the Capitol and the White House. A few months after the victory at Saukenuk, the British decided to seize New Orleans. "I have it much at heart to give them a complete drubbing before Peace is made," the vice-admiral who commanded the British forces in the gulf wrote. He figured that his victory would ensure that "the Command of the Mississippi [would be] wrested from them."

The U.S. forces were led by a lean and angry soldier named Andrew Jackson, who upon his death in 1845 was heralded as the embodiment of "the true spirit of his nation." Technically, that spirit had been forged east of the mountains, where Jackson's father, an immigrant from Ireland, had worked himself to death trying to eke a living out of a Carolina farm. After his family's home was captured during the Revolution, fourteen-year-old Jackson refused to polish a British officer's shoes. This act of resistance earned him a sword-slashed scar on his head and hand. Thereafter, it seems, Jackson was a man of righteous anger; any slight sparked furious indignation. He followed no law but what he felt was right. He was, then, spiritual kin to Mike Fink, only dressed in better clothes—and he followed Fink's example, joining the rush onto the western waters.

Jackson worked on keelboats and flatboats, carting swan skins, feathers, pork, and beef—and notably, human slaves. Eventually he established a business empire in Nashville, along the Cumberland River, that included a tavern, a racetrack, and—since some of his customers paid in bartered goods—a trading depot that could carry the wares downstream to market. He entered politics, too, and in 1812 was a prominent enough man to be put in charge of the local militia. He declared to his troops that their greatest duty as Westerners would be to defend their mother river against invasion. After he was told his services were unneeded in New Orleans, he spent the next few years fighting the Muscogee in Alabama, rising to become a major general. In late 1814, with the long-feared invasion imminent, he triumphantly marched his warriors into the delta.

Jackson's men were mostly the hardscrabble sort who'd established farms along the western rivers—his fellow flatboaters, in other words. In New Orleans, they were joined by French-descended pirates, Choctaw soldiers, and free men of color. This motley assembly delivered a surprising ending to the war: they routed the royal army. The Brits were perhaps too well trained. As they streamed across a fallow field of sugarcane just downstream of the city, they refused to abandon their orderly lines—even as they were met by a constant barrage of musket fire. A quarter of the eight thousand British soldiers died, compared to less than a hundred U.S. casualties.

For the Americans, the Battle of New Orleans was a triumph after years of chaos and loss—enough of a triumph, apparently, to finally settle the issue of who owned the river at the continent's heart. After the battle, the British mostly abandoned their Indigenous allies, which helped ensure that the length of the river would fall into secure U.S. control. The army sent a force north to build a fort at Rock Island; it dispatched Prospect Robbins and Joseph Brown west, on their trek through the Arkansas swamps. The imagined grid of the empire for liberty was, chain by chain, being laid atop the land.

Within a few years, January 8, the date of the Battle of New Orleans, became a widely celebrated American holiday—a second Independence Day. The massive British death toll was extolled in newspaper poetry. When Andrew Jackson ran for president thirteen years after his victory, he chose as a campaign song an old ballad that celebrated the battle. By then, the history was already being warped. Despite the diverse army that defended the city—"perhaps the most racially varied 'American' military force ever," according to scholar Thomas Ruys Smith—the song reduced the New Orleans battalion to one key identity: "ev'ry man was half a horse," the lyrics claimed, "and half an alligator." Rather than a nation of cultivators, the watershed had spawned an empire of aspiring Mike Finks.

The relict Mississippi River channel near Kaskaskia (Randolph County, Illinois).

5

The Office of River Improvement

The brief and glorious steamboat era

To the crowd assembled along the Tennessee riverbank in 1829, Henry Shreve's steering must have appeared bold, if not suicidal. His new boat, the *Heliopolis,* was aimed directly at a massive log that clawed up from the water—the sort of tree that was a boat wrecker, a death dealer. The *Heliopolis* was in a sense two boats, each with its own steam engine, their two platforms joined by a thick iron-plated bar. Shreve aimed that bar at the great snag.

The collision was thunderous. The boat jolted fiercely. The snag popped free.

This little removal was a great step forward. The first waves of Americans found that millions of years of geology had built something spectacular along the western rivers—something useful. But construction had been sloppy, unsuitable for the emerging dream of an empire. Now Shreve was fixing nature's mistakes. Men dragged the tree aboard the *Heliopolis* with winches, then sawed the wood apart. They dropped the root ball into the water, where it sank to the bottom. The timber they floated to shore. A few decades earlier, in his guide *The Navigator*, Zadok Cramer had called this stretch of water one of the most dangerous on the Mississippi, due to the thicket of snags beneath the water. Now, after eleven bone-rattling hours of labor, the danger was erased. The *Heliopolis* moved downstream to the mouth of the Arkansas River and over the next seven months cleared more than fifteen hundred snags.

Shreve left behind a few letters and reports, but these are mostly official correspondence, nothing that offers much scent of his personality. The only contemporaneous portrait I've been able to find was drawn in charcoal and chalk late in Shreve's life. He is puffy and white-haired, with the same stern and self-satisfied look that presidents feature on coins. In the absence of hard facts, his biography has become shrouded in doubtful legends. But Shreve deserves that look of satisfaction. Set loose with his snagboat, he likely did more to change the ecosystem of the antebellum South than any other individual—bashing open Mississippi's ancient waterways so thoroughly that, in a strange way, they began to disappear.

The first steamboat arrived on the Mississippi in 1811, traveling a river in chaos. Early that year the slave rebellion had disquieted New Orleans; now the war with Britain loomed. Even nature seemed unsettled: squirrels were said to have migrated south, "pressing forward by tens of thousands in a deep and solid phalanx." Then, not long after the steamboat *New Orleans* curled out of the Ohio River and onto the Lower Mississippi, a series of earthquakes rocked the frontier.

These quakes remain the most powerful to strike the United States east of the Rocky Mountains.* Fissures cracked open, swallowing buildings.

* The location of the earthquake's epicenter, in Missouri's bootheel, at the top of the Mississippi Embayment, has everything to do with the river's geology. Hundreds of millions of years after its creation, the Reelfoot Rift continues to slip and slide, causing an earthquake roughly every five hundred years. A 2009 report commissioned by the Federal Emergency Management Agency noted that given the lack of preparedness in the region, another quake the size of those of 1811 and 1812 could deliver hundreds of billions of dollars in damage.

Strips of forest slipped six feet into the ground. Stream bottoms thrust upward, and the water that spilled across the surrounding terrain was, despite the winter chill, "over blood heat." The ground erupted into volcanoes of mud and water and sand. Entire riverside towns were left in ruins; church bells rang as far afield as Charleston. According to one account, men on shore begged to come aboard the *New Orleans*, but the boat had no space for refugees. Other frontiersmen seemed to fear the noisy, smoke-spewing contraption more than the suddenly unsteady ground.

Even the crew seemed to be in awe of their accomplishment. "One of the peculiar characteristics of the voyage was the silence that prevailed on board," John Latrobe wrote decades later, based on stories that his eldest sister, a passenger, told. Another anecdote in Latrobe's account seems almost too perfectly symbolic: when the boat rounded the bend at Cairo, finally reaching the Mississippi River, a Chickasaw canoe appeared out of the flooded forests, intent on a race. There were as many paddlers in the canoe as there were crewmembers on the *New Orleans*. For a time, their human engine kept pace. They could not last. According to Latrobe, "the Indians with wild shouts, which might have been shouts of defiance, gave up the pursuit, and turned into the forest from whence they had emerged." The old ways disappeared, eclipsed by the arrival of new technology. The problem with this fable is that the *New Orleans* hardly changed the frontier.

The boat had been built that spring in Pittsburgh, out of white pine floated down the tributaries and iron forged in local plants. The most essential component, the hundred-ton engine, had been built in New York and hauled in pieces through the mountains. Don't picture the kind of lavish wedding cake of a vessel that later became so famous on the Mississippi. The *New Orleans* was drab and utilitarian; with its curved hull, it looked like an oversize keelboat—though at twice the typical dimensions, it was able to carry double the cargo and still fit eighty passengers on deck. The two paddlewheels did not lie at the boat's rear, as on later boats, but clasped its center. A smokestack emerged from the deck, along with two ungainly sailing masts, in case the pilot had to rely on the wind. The one nod to aesthetics was a coat of sky-blue paint.

The *New Orleans* embarked on its maiden voyage in October and, despite the quakes, reached its namesake city by January. In later editions of *The Navigator,* Zadok Cramer reported that in its first year, the boat returned a $20,000 profit—a 50 percent return. The boat was a financial windfall, then, though its owner, Robert Fulton, ran the *New Orleans* back north only as far as Natchez, fearing the treacherous shoals upstream. Nor did the rest of Fulton's fleet fare much better. His next western steamboat, the *Vesuvius,* was intended to establish regular trade between New Orleans and Kentucky. On its first trip the boat ran against a sandbar, where it sat stranded for six months. By the time the *Vesuvius* was freed, the *New Orleans* had sunk while moored overnight near Baton Rouge. The river dropped a foot and half, and the steamboat was impaled on a stump that lay hidden beneath its waters.

Louisiana's legislature had granted Fulton an eighteen-year monopoly, the exclusive right to run steamboats throughout the territory. But Fulton's rivals must have sensed his weakness. The first boat to challenge the law, the *Comet,* launched from Pittsburgh in 1813, before Fulton even got the *Vesuvius* on the water. Next came the *Enterprise,* piloted by the mysterious Mr. Shreve.

We do know enough about Shreve to be able to sketch his early life. His father, Israel, served in the Revolutionary army as a colonel, which displeased the family's New Jersey Quaker community. So in 1788, when Henry was two, the Shreves headed west into a valley still tangled in international intrigue. Israel joined an expedition that was headed downriver to survey the site of a potential village that disaffected Americans were hoping to establish in Spanish Louisiana. The project fizzled; the Shreves instead settled on a property on the banks of the Washington Run, which Israel purchased from the nation's first president.

This little creek was a tributary of the Youghiogheny, which runs into the Monongahela, which merges with the Allegheny in Pittsburgh to form the Ohio River. In 1807, as a twenty-two-year-old, Shreve followed these waterways, leading a crew to St. Louis, poling a vessel he had built himself. Three years later Shreve traveled even farther, pushing up the Mississippi to reach the Fever River, where he acquired sixty tons of lead from the Sauk. A rush of traders followed, but Shreve himself never bothered to return. He seemed restless, always seeking new adventures. Soon after the *New Orleans* launched, a team of investors in Pennsylvania tapped the young riverman to guide their new boat, the *Enterprise*, down the dangerous river into New Orleans.

The *Enterprise* arrived in early 1815, bearing a cargo of munitions. Andrew Jackson had foiled the British invasion the day before and remained nervous about another attack. So he commandeered the steamboat, ordering Shreve north to Natchez, where a set of keelboats loaded with ammunition had been stranded. Legends suggest Shreve spent the next few months flouting the Fulton monopoly, demonstrating his boats' ease on the western rivers—zipping south to the river's mouth, then hundreds of miles upstream, where he headed on to the Red. Finally, in May, when Shreve announced his intention to turn back home, Fulton's outfit flexed its muscle and had him arrested for violating its monopoly. Shreve was ready: he'd hired one of the city's best lawyers, who bailed him out. The pilot left immediately, allowing his lawyer to sort out the lawsuit. The ruling has been lost to history, but Shreve's accomplishment is not: twenty-five days later, the *Enterprise* arrived in Louisville, becoming the first boat to defeat the current of the mighty river.

Fulton's company soon offered Shreve a half interest in its monopoly. Shreve declined, believing, perhaps, in Jefferson's natural law: the river should be free to all. When he returned to New Orleans in a new boat in 1817, Shreve stayed on board the *Washington* so the city marshal would have to arrest him in front of a riverside crowd. The sight of Shreve in chains set off an uproar. Even better, the judge dismissed the case for lack of jurisdiction.

Already seventeen steamboats were defying the law, servicing New Orleans. Many more worked the rivers upstream. The monopoly was soon abandoned, and Fulton's company turned its focus to the smaller rivers in the northeast, offloading two of its three Louisiana vessels. The western waters—nearly half of a continent—were open for business.

Which is not to say the continent was easy to traverse. When the *Virginia* became the first boat to reach St. Paul, in 1823, the trek north from St. Louis took twenty days. At one point, a passenger disembarked to take a walk while the ship was loading wood. He missed departure but still managed to catch up to the boat, which was stuck on a sandbar, by walking upstream. If the bars were a headache, the rocks and rapids could be mortal hazards. In Iowa, the overloaded *Virginia* scraped against exposed bedrock and narrowly avoided catastrophe. The ship had to drop some cargo and try again.

The Supreme Court officially declared Fulton's steamboat monopoly illegal in 1824: the waterways, as paths between the states, were to be overseen by the federal government. Perhaps because the ruling focused on water, it's rarely noted that this was a great leap forward for the nation's public property. Decades before anyone dreamed of a national park, the Court affirmed our rivers as common holdings—as something that could not be claimed or sold. Not that they'd be protected for their beauty. No, now that they were federal property, the rivers were ready for, as it was known in the era, their improvement.

Who would do the improving? One candidate was the nation's small cohort of military engineers. During the Revolution, talent to build almost anything, fortifications especially, had to be imported from France. In 1803 Thomas Jefferson launched a "corps" of engineers as part of the new military academy at West Point, New York: future officers would be trained in mathematics, surveying, and hydraulics—a way to ensure a steady future supply of military engineers. Over the next two decades, this corps turned from a set of military instructors into an elite wing within the army's engineering department that oversaw many federal construction projects. After all, no one else in the country was equipped to build roads and canals. Now in 1824, after the rivers became public property, Congress gave the military engineers $75,000 to improve navigation on the Mississippi and the Ohio.

One key task was to deepen the river. For a first attempt, an army engineer borrowed a solution that his French colleagues had tried on the Loire.

A crew working from flatboats installed timber pilings to form a wing dam, a wall in the water that jutted from the right bank at a 45-degree angle, pointing downstream. This dam, built near Louisville, focused the Ohio River's flow; the faster and more forceful current blasted through a troublesome rock bar. This yielded a river that, in this one spot, at least, remained four feet deep year round. The dam stayed in place for nearly fifty years before it was repaired and extended.

Congress also instructed the Corps of Engineers to remove the snags, but for this problem the military had no ready solution. So in newspapers across the country, the agency's chief advertised a thousand-dollar prize to the man with the best idea. The winning proposal came from a steamboat captain who said he would build a pair of parallel flatboats. Together, the boats supported a winch with an iron claw that could be attached to a snag; the crew, turning a capstan, could drag the wood out of the water.

This device barely worked, and by 1827, Congress instructed the Corps to find someone else to lead the snag-pulling project. The obvious answer was the hero of the western rivers. Henry Shreve was put in charge of his own new subagency, the Office of Western River Improvements. And once he was named superintendent, Shreve seemed determined to find a way to improve the rivers.

If Shreve's approach lacked the grace of the French-inspired dam on the Ohio, perhaps that's because he was a closer match to the rough-and-tumble culture emerging on the river. When he was presented with an obstacle, Shreve just countered with whatever force was necessary to knock the thing loose. Near the mouth of the Ohio River, his crews used crowbars, drills, and tin boxes of gunpowder to remove two hundred boatloads of rock. Other teams undertook similar work on the Upper Mississippi, clearing a four-mile-long channel through the Des Moines rapids, a nasty snarl of water several hundred miles above St. Louis.

Shreve's breakthrough, his snagpuller, came by a simple reworking of the failed effort of a few years before. Shreve replaced the human-powered capstan with the high technology of the era: a steam engine. After his success in Tennessee, Westerners petitioned for more such boats—"Uncle Sam's Toothpullers," as they became known. Thus arrived the *Archimedes*, the *Eradicator*, and the self-aggrandizing *Henry M. Shreve*. Under the captain's direction, the snagpullers fanned up the major tributaries; soon insurance rates for shipping dropped by half. The workers on these crews were civilians, but the Corps of Engineers noted that their service was nearly equivalent to combat duty. Disease was rampant, drownings common. Men were sometimes maimed after tangling themselves in the snagboats' ropes and chains.

To prevent more snags from falling into the river, Shreve cleared the banks, too, removing nearly 75,000 trees in just three years. The hundreds of wood yards that dotted the riverbanks, where entrepreneurs sold fuel to the passing boats, may have felled even more of the old forests. A small steamboat might require twenty-four cords each day; a large boat would burn through seventy-five—enough wood to build fifteen cabins, according to the calculations of archaeologist Terry Norris. Just the fleet of steamboats that was headquartered in St. Louis must have burned enough wood in a year to build seven hundred thousand such buildings.

The river had always meandered, but all this uprooting unleashed a new kind of wanderlust. Without roots to hold the soil in place, the banks weakened. The channel shifted and often widened, expanding by 50 percent in the stretch between St. Louis and Cairo in the four decades after Shreve went to work. Norris believes that four of the seven villages that the French had cleared in the American Bottom were buried under the resulting tides of mud. Kaskaskia, one of the river's earliest colonial settlements and the site where George Rogers Clark established his headquarters during the Revolutionary War, had briefly served as the territorial capital of Illinois. In 1881, after years of flooding, the river forged a new path to the east of the village. Now, unless you're traveling by boat, it's impossible to reach Kaskaskia from the rest of Illinois; the village lies on the Missouri side of the river. Not that many visitors bother: perhaps a dozen people still live in Kaskaskia, which consists of little more than a crossroads surrounded by low, wet fields. Ninety-five percent of the original village site was destroyed as the channel shifted through the years.

The scope of Shreve's ambition is made clear by his effort to shorten the Mississippi's famous horseshoe bends. He retrofitted the *Heliopolis* into a dredger, attaching oversize boxes to pulleyed cables that were powered by the engines. Then in 1831 he steamed south to a long curl of river in Louisiana.

In earlier eras, when travel was conducted mostly by canoe, the common choice was to portage a few miles here rather than traverse the thirty-seven miles of the bend. For a steamboat such portage was impossible. So after dropping casks filled with explosives into twenty-foot-deep pits, Shreve positioned his snagboat at the southern edge of the bend's narrow neck. Using his boxes as shovels, he began to tear out the loosened mud. After two weeks, he had dug a trough that was seventeen feet wide. Then he broke through to the far side of the bend.

The water surged. Within days, the canal was a half-mile across and as deep as the main channel. It *was* the main channel. But the results proved troublesome: the mouth of the Red River, which poured its water into the bend that Shreve had abandoned, grew choked with shoals.

Fortunately for Shreve's reputation, the Red later became the site of his greatest triumph. Ambitious planters had long seen promise in the land along its banks, but there was no passable channel. The loose soils upstream in its watershed, in the southern Great Plains, easily shed their trees, which after centuries had accumulated into series of logjams that plugged 150 miles of the river. What everyone called The "Great Raft"—a name that seems rather small for something so geologically monstrous—could grow in a single year by a mile, as new trees were swept downstream. During early cattle drives into Mexico, cowboys walked across with no fear of drowning. Shreve noted willows and cottonwoods sprouting from the older, rotten wood. The trunks stood as broad as eighteen inches—a forest growing atop a river. And all this wood trapped the river's mud, too, creating shoals that pushed the water outward. Thus the Red River was insulated on both sides by lakes and bayous and wetlands, a jungle of swamp and water six miles wide.

In 1833 Shreve dispatched the *Archimedes* to the foot of the raft. Working with four steamboats and a crew of 150, he slowly extracted the wood. The work was brutal, the heat unrelenting. The only way the crew could shelter from the hordes of insects was to dive into the water. Still, in just ten weeks Shreve managed to clear seventy miles. The Red now flowed at three miles an hour—a seemingly laggard pace, but per Shreve's estimates, more than ten times faster than before. Late the next year, Shreve returned with double the crew. He worked from December through May, but because they'd already ripped through the oldest, rotten wood, the work was harder. When the season ended, twenty-three miles of raft remained. The river was not fully cleared until March 1838, and when the river rose that June, more trees piled up in their place. This new raft was cleared a few years later, only to reappear within weeks. Even so, the territory changed completely. As the raft shrank, upstream bayous drained more quickly. A rich mosaic of sloughs and swamps that reached far north as Arkansas went dry.

As speculators arrived, hoping to grab the best of the territory, the federal government paid the Caddo Nation to leave. The money and gifts were scheduled to arrive in installments; the Caddo found that one shipment, which was supposed to equal $10,000, consisted of thirty blankets and a few trinkets. The federal government, meanwhile, turned around and sold the land, earning $2.7 million by 1837. Shreve himself, alongside six partners, got in on the action, buying a tract of land atop a bluff that overlooked the upstream edge of the raft for $5,000.

Shreve's investment is a reminder of how thoroughly his project of river-clearing was entangled in the nation's imperial mission. The easier it was to travel, the farther the empire could spread. The U.S. government was already foisting treaties onto other peoples along the Arkansas and Missouri, seizing land along every major western tributary. The most effective strategies—which generally involved exploiting old political tensions, working outside of the typical tribal channels, and dangling threats of violence—had been honed by Andrew Jackson.

The general spent the decade after his New Orleans victory leading armies across the Southeast, securing land for white settlement. And this accomplishment was well received, at least within the watershed: when he ran for president in 1828, Jackson carried every state that touched the western rivers. As president, Jackson's signature piece of legislation was the Indian Removal Act. Even at the time, one congressman called that a "soft" name, given the inherent violence. In Mississippi alone, there were 23,000 Indigenous people, more than in all the northern states combined, and Jackson intended to force them—along with the rest of the Indigenous people left in the Southeast—across the Mississippi, into land the settlers had not yet decided to swallow.

Alexis de Tocqueville encountered some of the first deportees in Memphis in 1831. "The cold was unusually severe," he wrote. "The snow had frozen hard upon the ground, and the river was drifting huge masses of ice." Many of the Choctaw were wounded and sickly, but they were silent as they boarded the steamboats. Their dogs were not permitted to follow, and as the boats departed, the animals "set up a dismal howl, and, plunging all together into the icy waters of the Mississippi, they swam after the boat." Conditions proved even worse where the boat ride ended, at Arkansas Post, the small village founded more than a century earlier by La Salle's lieutenant near the mouth of the Arkansas River. Several thousand Choctaw had nowhere to camp except the frozen riverbank. Their tents offered meager shelter against temperatures that hovered in the teens. Over the next few years, as fifteen thousand Choctaw headed west, at least 2,500 died.

Back in the Choctaw homeland, the surveyors and bureaucrats scrambled to prepare the land for auction. The western edge of the tribe's territory reached into the Mississippi River's floodplain, though this was a remote region where some Choctaw, far from the leadership of the nation, were already beginning to shed old traditions and assimilate into American society. They'd largely adopted Western dress and agricultural techniques—becoming, in essence, the embodiment of Jefferson's cultivators. Now thousands of Choctaw opted to stay on their land and become U.S. citizens, as was allowed by the treaty. But thanks to incompetence—if not willful fraud—most of their property was sold, too. The historian Claudio Saunt has calculated that in total the Choctaw lost as much as $10 million, which at the time would have been enough to capitalize one of the country's largest corporations.

Andrew Jackson, now president, had sent trusted allies to negotiate the treaty with the Choctaw; he was eager to get the land sold quickly, to further spread the white empire. He liked to talk about his devotion to Western frontiersmen, and by now "preemption laws" were in place, intended to protect the small landowners who might become Jefferson's cultivators. Squatters who improved a plot before it was put up for auction were allowed to buy the land at the lowest legal price. If the Land Office felt the need to move the squatters, it sometimes bought them out with "floating" grants to any quarter-section of land in the survey. These bought-out squatters selected the best possible plots, but rarely managed to pay even the minimal fees required to retain them. Despite the laws' intentions, then, "the rich are the persons benefited in the end," as one surveyor noted, "because the poor cannot pay for their land, and all they can do is sell their claims and remove to some other place."

The rise of steamboats helped lure these wealthy buyers: they wouldn't have to focus on raising crops that could be bartered locally. They could switch to a monoculture. And they were learning just which forms of monoculture paid off. To the south, in the delta, an inventive farmer had learned how to granulate sugar for commercial sale, which soon became the key crop in Louisiana. Around Natchez, meanwhile, wealthy men, including the part-time explorer William Dunbar, became intrigued by cotton, whose price had soared in recent years. Soon the Yazoo Basin became famous as the best possible ground for growing the world's most important commodity. Still, since the southern bluffs were safer from floods, they wound up being settled first.

Frederick Law Olmsted, the landscape architect who would go on to design Central Park, passed through the country near Natchez the 1850s. He asked one resident whether there were any poor people around. "Of course not, sir," the man replied. "Every inch of the land bought up by swell-heads on purpose to keep them away." Some of the swell-heads did not actually live on their plantations. "Must have ice for their wine, you see," the man explained, "or they'd die." They chose to live in the urban centers, in Natchez and New Orleans. In summertime, many absconded north to New York, a better place to flaunt their wealth. Those who did live alongside their fields often occupied opulent mansions. Greenwood

Leflore, for example, was a half-white Choctaw chief who helped sell the tribe's Mississippi homeland, then stayed behind himself, overseeing hundreds of enslaved workers who raised his cotton. In the 1850s he filled his home with hickory furniture that he custom-ordered from France, overlaid with gold and upholstered in crimson silk damask. The house had fifteen rooms and six fireplaces and could accommodate two hundred guests.

Slavery had been declining at the end of the eighteenth century, but this new rush of wealth into the South helped revive the market. Because malaria was rampant, planters grew accustomed to "seasoning" their workers, waiting out a long period of illness after each batch arrived in the region. Inevitably, the disease killed some slaves, but this was the cost of business: common wisdom suggested that by the time he died, a slave was likely to have earned his owner enough money to pay for two replacements.

Only the boldest of the planters would venture down into the river's bottomlands, where the swamp forest was so thick that simply to walk, one had to hack open a trail with cane knives and axes. These planters also tended to be the wealthiest, able to afford the workforce needed to make this landscape work. In 1830 in Washington County, the first center of American settlement in the Yazoo Basin, more than half the population was enslaved. Two decades later the ratio of enslaved Black residents to whites had risen to more than 14 to 1. Eventually, one planter in Issaquena County owned more than a thousand slaves, an accumulation of human wealth that made him a millionaire. Before the Civil War, the county was the nation's second richest when measured per capita. In the wealth-per-person ratio, however, most of Issaquena's residents were not then counted as people. In the years since its Black population was officially flipped from the ratio's denominator to its numerator, Issaquena has consistently ranked among the poorest places in the country.

The challenge of the landscape meant that the floodplain's interior remained mostly untouched. These first plantations tended to run along the Mississippi itself, or along the Yazoo, or they perched on the natural levees on the banks of oxbow lakes, where the enslaved workers hacked away at the ancient forests. Sometimes the trees were burned, and in other cases they were "girdled," the bark peeled back in rings to expose the trunk to insects and rot. Then the first cotton plants were set amid ghost forests—fields of dying trunks. Thus, as one magazine correspondent put it, the settlers "beautified the wilderness."

The rows were precise, spaced three to five feet apart. The furrows lay three inches deep and often angled east to west to ensure maximum exposure to sunlight. The simple shape of this landscape made it easier to surveil the workers. Men on horseback loomed above the cotton, watching the men and women and children as they stooped, plucked the bolls, and stuffed the lint into their sacks. Once the cotton was ginned and bundled, it was loaded onto a steamboat to be sent downstream and shipped overseas.

On the Upper Mississippi, the steamboats carried a different cargo. In the wake of Henry Shreve's successful trade, tens of thousands of pounds of lead were being stripped out of the Sauk's old mines. Eager men were arriving by the thousands in Illinois, where the hills grew cluttered with hastily assembled smelting furnaces. These men were known as "suckers," since, like the fish of the same name, they'd head north in the warm months, then back south when the rivers began to freeze. These were a different sort from the swell-heads along the southern river: "thievish, poor, dirty, low-lived, rough scruffs," as one sniffy Illinoisian complained.

The Sauk, with their tidy farming village, were closer to the Jeffersonian dream than these suckers. Despite the 1804 treaty, the tribe had never abandoned their homeland. But their time there was drawing to a close: the suckers noticed Saukenuk in 1828—empty, since the tribe was gone for the winter hunt—and liked what they found. A handful of miners began to squat in the village, hoping to give up the hard work of lead smelting for a more settled life. Black Hawk was in his sixties now, but he was still committed to the ancestors who lay buried at Saukenuk. He became the de facto leader of the recalcitrant Sauk who refused to leave their home. Even after the federal government auctioned the land, he kept leading his followers back each spring, after hunting season, to live in an uneasy existence with the white squatters.

By the summer of 1831—at the same time the government was preparing to "remove" the Choctaw—the Illinois militia was gathering soldiers to enforce the Sauk eviction. Aware of the threat, Black Hawk and his followers slipped out of the village. The soldiers seemed miffed at missing their chance for bloodshed: they burned the empty lodges, then unearthed Sauk corpses and burned those, too. If anything, the desecration intensified Black Hawk's commitment. He crossed the Mississippi again the next spring, leading several hundred soldiers from various tribes, plus a hundred women and children and elders. The group—the British Band, as they were known, for the flag they flew, a symbol of resistance—had no clear destination. But Black Hawk made a point to show no fear, instructing his followers to sing and beat their drums when they knew the U.S. military drew near. Eventually, the inevitable battle erupted. After an early victory against a company of drunken Illinois rangers, Black Hawk's group headed north. They hid out for most of the summer in the tangled marshland of Wisconsin before they were discovered and fled through the forests, toward the great river.

Finally, on August 2, along the banks of the Mississippi, U.S. soldiers finished this little war. The British Band—starved and disease-racked after months in the difficult landscape—"fell like grass before the scythe," as a federal Indian agent later reported. Mothers swimming toward safety, carrying infants on their backs, were not spared. The river changed color, the agent reported, "tainted with the blood of the Indians." A few decades later, when a new settler town was established near the site, it was given the name of Victory.

Black Hawk survived, abandoning his followers and hiding out with a small party in a camp just to the north of the site of the massacre. The local tribes wanted no trouble, so when they found Black Hawk, they forced him to turn himself in. He spent the winter in chains inside barracks south of St. Louis. Then the U.S. government sent him east on a tour that was intended to demonstrate its great size and power. Andrew Jackson, when he met the war chief, underlined the point: "Our young men are as numerous as the leaves in the woods. What can you do against us?"

When the tour moved through western towns, where the Indigenous threat remained palpable, the crowds were often angry. In Detroit, effigies meant to represent Black Hawk were hung and burned. But to the east, gentlemen asked to shake the hand of the old warrior. Ladies kissed his cheek. Perhaps Black Hawk was embraced because easterners were eager to absolve themselves of the tragedy unfolding along the Mississippi River—because some people remained uncomfortable with the half-alligator American ways.

Not that the policy changed. The U.S. government used Black Hawk's truculence as an excuse to take six million acres in Iowa from the Sauk and their allies, the Meskwaki, despite the fact that British Band had acted against the wishes of many chiefs. Many accounts suggest that Black Hawk, meanwhile, crumpled into hollowness. After years of refusing American ways, he began guzzling whiskey and wearing broadcloth shirts. He strung medals over his chest and carried a silver-headed cane, a gift from a senator. He'd lived a life of trauma and watched so many of his kin be killed. Acquiesce or die: such was his choice.

The American expansion was relentless. The treaties were gobbling up more and more Indigenous land—up the western tributaries, and north into Minnesota. The country here remained rustic; its few white residents were fur traders and, increasingly, loggers with the timber companies that supplied the lumber needed by settlers advancing ever farther west. The year after Black Hawk died, in 1839, many Ojibwe gathered to watch the first steamboat ascend the St. Croix River, which forms the border between Minnesota and Wisconsin. The tribe expected the much-needed foods and supplies they'd been promised in exchange for land. Instead, the boat spilled out workmen and equipment, arrived to construct the first sawmill along this tributary.

Even today, steamboats remain the river's most familiar emblem. You'll find them, or at least their replicas, docked at the levee in New Orleans, or stopped over in Natchez and the small forgotten villages upstream. As they await passengers for jazz dinners and overnight larks, they display familiar ornamentations: elaborate filigrees winding along the railings; chimney tops carefully cut into the shape of crowns. Chandeliers adorn the ballrooms, and calliopes bellow cartoonish old songs. In their heyday, such decorations served a real function. On any given afternoon, a hundred or more steamboats might be tied up along the levee in Cincinnati or New Orleans. Beauty was how you stood out.

Even far from the river, the world was fascinated with these vessels. Engineers arrived from India to study their design. The river, meanwhile, went on a tour of its own: for a time in the 1840s, the hottest ticket in the world's major cities was to Mississippi panorama shows. Only a few feet of these paintings were displayed at a time, so as the canvas was unwound from one roller and respooled onto another, viewers were able to approximate the experience of riverboat travel. The panoramas featured the river's great earthworks and its Indigenous people—some "reclining lazily" in camp, according to an accompanying pamphlet, some dancing their war dance, some hunting bison. Steamboats featured prominently, too. Steamboats loading firewood. Steamboats racing upriver. Steamboats in port. The remnants of sunken steamboats, reaching up from the muddy flood. At one show, the program assured attendees that these were the "correct likenesses of boats that are now plying on those waters": the *Uncle Sam,* the *Peytona,* the *West Wind.*

There were reasons to be wary of these vessels. *Lloyd's Steamboat Directory and Disasters on the Western Waters,* published in 1856, purported to offer "a copious detail of those awful and distressing accidents which have been of too frequent occurrence." Its index serves as "a sort of nightmare poem of alphabetized Americana," the historian Walter Johnson says: "*America,* explosion, of; *America South,* burning of; *Anglo Norman,* explosion of; *Atlantic* and *Ogdensburg,* collision of." On and on, for pages. This directory was a work of muckraking, exposing the "gross and criminal mismanagement of steam power," and it became a best seller. But the dangers hardly seemed to matter compared to what the steamboats could accomplish.

Steamboats delivered cotton bales to port. Steamboats towed the logs that were being ripped from the northern watershed. Steamboats carted enslaved Black families to their conscription amid the new floodplain farms. Overpacked, pulsing with cholera, steamboats facilitated the deportation of tens of thousands of Indigenous people. One magazine writer in 1848 summarized their impact: the introduction of steamboats into the western waters, he wrote, "has contributed more than any other single cause, perhaps more than all other causes which have grown out of human skill, combined, to advance the prosperity of the West." A newspaper noted that New Orleans had been "brought within four days travel of St. Louis." We're so used to speed today—to hopping across the continent in a single afternoon—that it's hard to conceive of the shock of this revolution. The steamboat was the great talk of the era, eventually birthing a cliché: distance had been annihilated by steam.

The states and territories along the western rivers were no longer a frontier; soon they'd house half the country's citizens. Cincinnati, home to a few thousand people in 1810, had become a city of more than a hundred thousand, its streets lined with grisly hog-packing factories. Pittsburgh had grown tenfold. The steamboats coursing north into the newly established states in the upper watershed carried more settlers than the famous gold rush luring prospectors west to California. To the south, New Orleans had become the largest city in what was still then known as "the West"—the third-largest city in the nation.

A certain flavor of wildness waned with all this growth: it rendered the keelboats irrelevant, and with them went the drunken boatmen. (As for old Mike Fink, he joined an expedition up the Missouri in the 1820s, perhaps seeking a landscape still untamed and mysterious. Soon he was dead, shot in a drunken dispute. His passing doubles as a poignant marker: the next year the Corps of Engineers began its program of improvements.) Only the flatboats remained. The oversize rafts still offered a cheap means for carting goods to market. The crews mostly consisted of young white men, farm boys from the emerging Corn Belt. They faced little danger. No pirates or raiding Shawnee, not even snags. The only real peril was the homesick blues. Still, these boys self-consciously recalled the river's half-alligator days. To amuse themselves, they howled back and forth between boats like wolves. "There is something more romantic than you can imagine," one Terre Haute youth wrote in his journal, "in breakfasting on the deck of a broadhorn on a fine morning floating down the river." But the romance did not last: by the time he reached Memphis, this diarist was so sick of the heat and mosquitoes that he splurged on a room in the Exchange Hotel.

The boys usually rode home by steamboat, though often they could afford only the cheapest fares. So they slept in the lower decks, where the wise passengers eschewed the lice-ridden bunks and curled up atop the freight. For those on the decks above, the steamboat was a fashionable adventure, tame enough now for respectable ladies, who could see the river without ever setting foot on the muddy banks in the squalid little towns. The boats and all their grand ornamentation allowed passengers to forget for a moment the half-finished country in the alluvial valley, still mostly wild and empty: the boat itself was, as Mark Twain pointed out, "a new and marvelous world." The river, meanwhile, was reduced to scenery, something you watched as you leaned across the railing. Or in the case of the panoramas, it was even less substantial than that: two thousand miles of travel replaced by two stirring hours in an auditorium.

Eventually, thousands, if not millions, of people saw these displays. Henry Wadsworth Longfellow was at work on a long poem about the swamp country in Louisiana when he saw an advertisement for a panorama show in Boston. "This comes very *à propos*," he wrote in his journal. When he published *Evangeline*, its evocative descriptions of the southern swamps helped make the poem a national sensation. But Longfellow had never seen his exotic setting. "The river comes to me instead of my going to the river," he wrote.

Still, by standing on the riverbanks themselves—as young Samuel Clemens often did in Hannibal, Missouri, in the 1850s—you could watch a golden age unfurling. There were more boats, fancier and faster, than ever before. To be allowed to pilot such a craft was to attain the highest form of prestige. Clemens—who would become famous under his pen name, Mark Twain*—fulfilled his boyhood dream of becoming a steamboat pilot in 1859. "I loved the profession far better than any I have followed since,"

* The name is drawn from the calls the steamboatmen used as they measured the

he later wrote. A pilot, he noted, was so fully in command of his vessel that he became "the only unfettered and entirely independent human being that lived in the earth"—a very American sort of accomplishment.

Back in 1848, a Corps of Engineers report had listed the miles available to pilots like Clemens: six hundred on the Arkansas; sixty on the Neosho, which reaches into Kansas; 375 on the Ouachita, the river explored by Jefferson's Scottish adventurers; thirty-five on the Salt River in Kentucky; three hundred on the Yellowstone River in Montana; two thousand on the Mississippi—and so on. Seemingly every backwater slough across the South had been pried open for the boats. This list, though, marked a high point. The steamboat had expanded to its limits, and it had no more rivers to reach.

Really, by Clemens's pilot days, anyone who looked beyond the glittered splendor to examine the numbers would have found an industry in decline. The steamboat altered the river valley, but once the steam engine was built into a locomotive, it helped the empire for liberty bust beyond the bounds of the watershed entirely. The upper river had had to close for ice each winter, while locomotives could push through winter snows. Straight lines of rail could pass through mountains, ignoring topographical constraints. So the nation's cargo found new routes. Over a few years in the 1840s, the amount of corn and pork delivered by river to New Orleans dropped by two-thirds. Even Tennessee cotton began to be diverted over the Appalachian Mountains, before being shipped abroad from ports in Charleston and Savannah.

The work to improve the river had begun to sputter, too. After 1837, a financial crisis cut back on the Corps of Engineers' capacities; soon most of their equipment for river work was mothballed or sold at auction. Shreve was removed as superintendent in 1841, after a new party took power, and in the decade that followed, navigation appropriations were sparse at best. When Shreve died in 1851, he was tangled in a feud with the government over who owned the patent to his snagboat. But by then even the consummate riverman was looking away from the Mississippi, investing in a rail line that he hoped would link his home in St. Louis to the Pacific.

---

channel, ensuring it was deep enough for travel. An archaic word for "two," *twain* signaled that the water lay two fathoms deep.

In 1856 the government officially shuttered the Office of River Improvement. And that same year at Davenport, Iowa, as if to mark the end of an era, the first train rolled across a bridge spanning the Mississippi. A longstanding geographical barrier had been breached.

Two weeks later the steamboat *Effie Alton* drifted into the bridge. Within minutes, it had burned to cinders. The bridge smoldered and collapsed the next day. Steamboats along the riverbank celebrated with cheers and whistles and bells, which led many to conclude that the disaster was a kamikaze mission.

The case wound up in court, and a lawyer named Abraham Lincoln assembled a defense for the bridge company. Decades earlier he'd been one of the young men of the river: he'd served in the militia that fought Black Hawk; he'd ridden flatboats south to New Orleans to sell cargoes of flour and potatoes and pork. Now Lincoln was a rising politician, representing the needs of a fast-growing state. The frontier countryside beyond the river, he noted, was being developed at a rate never before known in the history of the world. "Mr. Lincoln," *Rock Island Magazine* wrote in a summary of the trial, "claimed that rivers were to be crossed."

Perhaps not everyone was yet convinced: the jury wound up hung, and the judge dismissed the case. That meant the bridge, rebuilt already, remained intact. The nation was plundering on westward. The river, as it had been for Hernando de Soto's soldiers, was just an obstacle in its path.

Indiana Chute at the Falls of the Ohio (Clark County, Indiana).

6

Dancing at the Skirts of Congress

The ambitions—and rivalries—of the ascendant engineers

In 1855, the year before the Davenport bridge spanned the river, the prominent engineer Lewis Haupt did his best to imagine what life had been like for his poor suffering countrymen just a half-century earlier. Back then, there had been no railroads, not even local tramlines. There had been no fuel except wood, no telegraph, no municipal sewer lines. It seemed "scarcely credible," Haupt wrote, that such a backward place could have preceded the nation he now knew.

The United States was urbanizing and modernizing. No one could think the United States might remain a nation of small farmers, an agrarian empire for liberty. No, this was a place of technological might—of steel and steam and rising cities. With a building boom afoot, a discipline arrived to supply the needed builders. Rensselaer Polytechnic Institute launched the nation's first college of engineering in 1850; Harvard and Yale soon offered programs, too. Over the next two decades, seventy more engineering schools appeared.

For this new breed of young men—and in this era, they were all men—the work was more than a career. The label *engineer*, Haupt sniffed, should not be applied to someone who, say, simply kept the engine of a locomotive running down a track. These men saw themselves as the vessels through which reason and knowledge could flow, yielding a better world for everyone. They were the spreaders of civilization. The task went beyond improving nature, then; the engineers had to confront the calcified political and social institutions that were holding back the world. On the Mississippi River, that would mean confronting the might of the army.

The Army Corps of Engineers had always been a strange fit for the young nation. During the War of 1812, as a small department, it proved its mettle by building forts and drawing up charts. But the focus on mathematics and theory at West Point—whose top graduates moved on into the Corps—set the military engineers apart from the rest of the era's culture. Rather than fussing around in textbooks, frontiersmen in young America got things done by plunging ahead and building *à la* Henry Shreve or by fighting tooth and nail in the style of Andrew Jackson. By 1830, the elite cadets at West Point were widely lambasted as fussy and bookish—their clothes too clean, their hats too feathered. The school, as an anonymous pamphlet put it, was home to "a privileged order of the very worst class, a military aristocracy."

Andrew Atkinson Humphreys serves as a fitting example. His father and great-grandfather were high-ranking naval officers; his biography, written by his son, traces their ancestry far enough up the tree to reach British royalty. Perhaps Humphreys inherited a sense that he should always get his way. As a child, he was impatient with schooling—with punishment especially—and for years had to be switched from tutor to tutor. Finally in 1827 his parents tapped family connections to install their sixteen-year-old hellion in the military academy.

Humphreys emerged from West Point in 1831 not quite disciplined—demerits left him thirteenth in his class, too low to immediately join the Corps of Engineers—but extraordinarily driven. He wanted to do "some substantial good," as he wrote in his diary a few years after graduation, while posted along the barren beaches in Cape Cod. The topographical work he was assigned there, conducting a survey of the Provincetown harbor, felt too simple, and to his disgust, his own self-guided program of study was constantly interrupted by less compelling tasks. Soon he at least got to see some military action, a campaign in Florida against the Seminole, though he contracted a fever in the swamps and was forced to resign.

After a few years as a civil engineer, Humphreys returned to the military, joining the Corps of Topographical Engineers as a first lieutenant. At the time, this agency was separate from the Corps of Engineers, though both bureaucracies earned the same dim reviews. The officers were known as dandies "dancing attendance at the skirts of Congressmen," as one Kentucky politician complained; they were "never seen in the hour of danger, and found only where favors were to be had."

Humphreys danced well. He could be charming—"very pleasant to deal with," as one government official said years later, "unless you were fighting against him, and then he was not so pleasant." He befriended congressmen and prominent scientists and jockeyed aggressively against his agency rivals for the assignments he desired. Still, at forty, as a captain, with no great feat to his name, the ambitious topographer must have felt stymied.

Then in 1849 the surging Mississippi River pressed through the levees upstream of New Orleans. Water lay atop the city for over a month, so much water that people talked of destroying New Orleans's own levee to allow the flood to drain. This disaster was bad enough that, in an era when river investment was dwindling, Congress commissioned a study to see what might be done to stop such floods.

"It is a work which I should desire," Humphreys wrote, "as it is one of much difficulty and great importance." So he danced again, cashing in on the favors he'd collected, leaning on family connections, and got himself the job.

The key question for Humphreys was how the levees that had been built along the river were affecting its behavior. What had started as a waist-high barricade along the New Orleans riverbank in 1720 had grown over the next eighty years into an embankment running nearly the length of the delta. After the Louisiana Purchase, the Americans noted the wall and copied it, extending the levees northward into the alluvial valley. Still, the system was plagued by gaps. The levees never managed to stop the endless floods.

Humphreys recruited three teams of engineers, providing each with extensive instructions. They conducted a topographical survey. They examined historical records of the "crevasses," the cracks in the levees blown open by the weight of the floods. They measured the river's cross sections and developed new methods to gauge its flow's velocity—not just at the surface but at various depths. The men, working in the riverside swamps through the hottest weeks of southern summers, fell sick for weeks at a time, delaying progress. Humphreys steamed down the river to keep an eye on the work, and he too grew ill—a case of "excessive mental exertion and intense application to business," a physician wrote. He had good reason to exert to himself: this work had become a competition.

Congress, by this point, was unsure whether the widely detested military engineers were the best choice for the job. Some of the funding that was meant for Humphreys wound up diverted to a separate project, to be completed by a civilian engineer. Charles Ellet was the rare American who had trained in Paris, at what was at the time the world's best engineering school. He was as ambitious as Humphreys but had already achieved fame—as much for flash and daring as for engineering work. A few years earlier Ellet carried himself across the Niagara Gorge in a basket hanging from a wire, the first ever crossing, and then built a rickety bridge of wooden planks and crossed that, too, in a horse-drawn carriage. Still, in terms of the story of the Mississippi River, Ellet is less a character than a foil. What matters is Humphreys's fury that someone else might impede on what he saw as his life's great work.

Ellet's report, based more on theory than on data, was finished in just six months, while Humphreys was still recuperating. Ellet concluded that building too many levees would lead to catastrophe. ("The process by which the country above is relieved, is that by which the country below is ruined," he wrote in his brief and elegant report.) Ellet suggested the nation needed instead a complex system of reservoirs to slow the floods, plus widened bayous in the delta—maybe even new artificial bayous—to help the floods get out to the ocean when they did arrive. The idea that levees alone might stop a flood, he said, was "a delusive hope, and most dangerous to indulge, because it encourages a false security." This astute plan, however, had one enormous problem: no one believed that the government had the authority to build such a system. The government could be involved in trade and commerce and transportation, but it was not in the business of keeping floods off private land.

Congress had at least settled on one compromise for the soggy southern riverlands: it decided to transfer the 9.5 million acres of federally owned swamps in Louisiana—untamed lands that no settler had bothered to buy at auction—to the state government, which would sell the land and use the profits to fund new engineering. Congress passed the Swamp Land Act in 1849, and within three years, Louisiana divvied itself into a set of "drainage districts" to oversee the construction of new levees and canals. Other states demanded their wetlands, too, and over the next decade, Congress passed similar laws on behalf of every state along the Mississippi River except Kentucky.

Once again, lawmakers had to reduce the complexity of the landscape to the kind of abstraction that could live within legal memos. If most of a plat was swampy, it was deemed a swamp. Accuracy was not a primary concern. In Mississippi, surveyors were paid based on how much swampland they were able to successfully identify. Rumor had it that one official pushed a canoe through the dry dirt of a piney forest, using the fact of its passage as proof of swampiness.

After Humphreys recovered from his illness, his superiors moved him to other assignments. But he still stewed over his lost opportunity, and in 1857 he convinced the army that he should dig the old Mississippi River data out of the archives to finish his long-delayed report. Despite Ellet's warnings, levees remained the de facto solution to flooding; by now, they stood three to four feet tall along the majority of the lower river's banks. But there were significant gaps, and local engineers were not always the best and the brightest; they built levees that were too narrow or too close to the river. The floods came too quickly, and too little could be accomplished, especially with malaria sweeping through the work camps. In the spring of 1858, a major flood smashed through the wall, destroying a fifth of the Louisiana sugar crop. The largest crevasse, a gap a thousand feet across, became a spectacle: an entrepreneur established a barroom across the river and organized boat tours to carry tourists up close to the damage. Clearly, there was still a problem to solve.

Occupied with other projects, Humphreys sent a young lieutenant, Henry Abbot, south to collect new measurements. (Humphreys later recognized this contribution by naming Abbot his co-author.) Then, in 1860, with the data in hand—and the nation on the verge of war—Humphreys sat for long nights as his desk, compiling a massive tome. The title alone was eighty-one words; for simplicity, it's mostly become known as the Delta Survey.* If you include the seven appendixes, the survey ran more than six hundred pages, all printed in a minuscule font. When the report was published, *The American Journal of Science and Arts* praised Humphreys and his team for their "unwearied industry and patient accuracy." The Delta Survey, the journal declared, was "one of the most profoundly scientific publications ever published by the U.S. Government."

Humphreys spoke with awe of the river, whose "enormous volume and apparently irresistible power impart to it something of sublimity"—a sublimity that was impossible to convey in words, he suggested. But words were not the point. This report was not a poem. The river was a collection of behaviors and phenomena that were technical and enumerable. Indeed, in his preamble to the report, Humphreys declared that his teams had ascertained "every important fact connected with the various

* In full: *Report on the Physics and Hydraulics of the Mississippi River Upon the Protection of the Alluvial Region Against Overflow and Upon the Deepening of the Mouths: Based Upon Surveys and Investigations Made Under the Acts of Congress Directing the Topographical and Hydrographical Survey of the Delta of the Mississippi River, with Such Investigations As Might Lead to Determine the Most Practicable Plan for Securing It From Inundation, and the Best Mode of Deepening the Channels at the Mouth of the River.*

physical conditions of the river and the laws uniting them." He used this information to develop a cumbersome formula that he believed could predict the behavior of any river in the world. (Within the decade, hydrologists would prove this equation inaccurate.) Ultimately, though, the point of the report was to determine the best response to floods. And Humphreys confirmed that despite the levees' poor performance so far, a bigger and better wall was the only viable approach.

This argument, in retrospect, seems grounded mostly in spite. Humphreys was prideful about how many measurements he'd collected, especially compared to Ellet's scant data. But many of his calculations wound up supporting Ellet's conclusions. They agreed on much of the hydrological theory, too. The trouble was that while Humphreys had "begun his survey with intellectual curiosity and honesty," as historian John Barry writes, "he had also always intended to write a masterpiece. No masterpiece can merely confirm another's findings." Once Humphreys was able to return to his task, he had made a point to dismiss his rival's measurements and calculations and suggestions. The complex engineering that Ellet recommended would be risky and costly, he said. Levees, by contrast, were cheap and simple, so they were the superior approach.

Not that it mattered, at least for the moment. A few months before Humphreys submitted his survey, South Carolina militiamen had fired on Fort Sumter. The nation was riven by war, and the levees were left to crumble and decay.

Like many well-connected West Point men, Humphreys rose quickly through the ranks, eventually becoming a major general. So the former wild child became the disciplinarian. His troops detested him, behind his back calling him "Old Goggle Eyes." He was known for his constant washing-up and his fastidiousness—the kind of pedantic attention to detail you might expect from a topographer.

Most of the extant photos of Humphreys come from this era. Sunken-cheeked and buttoned tightly into his uniform, he's never smiling. His mood seems to shift in a narrow range, from merely glum to smolderingly unhappy. His letters did describe one great, if rather perverse, source of joy: riding at the head of his division, leading his men into battle, he felt "like a young girl of sixteen at her first ball." It was as if he wanted to cast aside the old criticism that army bureaucrats shrank from scenes of danger. (He did not seem to care that this foolhardy charge killed more than a thousand of his soldiers.) In a farewell speech upon receiving a promotion, Humphreys declared that the war years fit him well. He was a dedicated surveyor, sure, but "anyone who knows me intimately, knows I had more of the solider than a man of science in me," he told the assembled troops.

The Civil War undid much of the work of the earlier eras. By the time the last treaties were signed, sunken warships littered the riverbed; dams and dikes lay buried under mud and sand. For some veterans, shell-shocked by the violence, the ragged river offered a refuge. They moved into houseboats assembled from driftwood and cheap pine boards and whatever other scrap the builders could find, then launched downstream. This tribe of "shanty-boaters," as they became known, grew to include descendants of the early pioneers from the Appalachians; they were in this era being pushed off their land by companies that had bought up the mountains to mine coal. An economic slump in 1875 sent more families out onto the water.

For the steamboats, the conflict had meant one final hurrah: after a brief bout of nervous quiet on the rivers, the pilots found themselves hired by the government to deliver soldiers and munitions, hogs and cattle and coal. They were praised as heroes, essential to the Union victory, though when the war ended, so too did almost all the work. The river had been closed to private shippers, who feared guerilla attacks, anyway. The war, then, finished the shift that had begun a few decades earlier: the railroads had become the primary path across the continent.

Perhaps it was patriotic fervor that led Congress to reboot the Office of Western River Improvements in 1866, as a gift to the steamboat heroes. Or perhaps it was just a sign of the times. Much of the South lay in ruins, and there was little local capital to rebuild, so the federal government took on the job. The whole Corps of Engineers was rebooted that year—detached from West Point and merged with the topographical division, where Humphreys had once served, to form a new powerhouse agency.

For its leader, Humphreys was the obvious choice. The surveyor and soldier extraordinaire became the army's chief engineer.

Soon the Corps would include a hundred officers, divided across multiple geographical districts. But even this staff could not keep up with Congress's new ambitions. Legislators tacked project after project onto omnibus river and harbor bills. So long as every legislator got something funded, no one was going to vote down a bloated bill. Sometimes the Corps's districts received more funding than they asked for. They built projects that they themselves had panned in their pre-construction analysis. But if Congress wanted it, why not get paid for the work? So the improvements threaded outward, touching nearly every tributary. As always, though, the biggest rivers got the most ambitious treatment.

A new snagboat headed downstream to rip the wreckage of sunken gunboats out of the navigation channel. Then construction commenced to replace the old wing dams. The raft of logs on the Red River was finally and completely removed in 1873 (the solution: nitroglycerin). To the north, Congress authorized Humphreys to survey the Upper Mississippi to determine how to keep its channel four feet deep. He delegated the task to one of the engineers who had helped complete his massive report, and by 1867, he had teams blasting out the rocks in the rapids near the mouth of the Des Moines River. The Corps was carving a canal around a second set of rapids, too, to the north, near Rock Island. The next year dredgeboats began to smash through the sandbars. But soon it became clear that this river remained beyond the army's power. New bars would appear so quickly that the engineers had to have a dredgeboat on call, ready to re-open the channel whenever a steamboat appeared.

To the east, on the Ohio, Humphreys worried over a bad set of rapids near Louisville, Kentucky, known as the Falls of the Ohio, where the river drops twenty-six feet over two miles. The "falls" were not a waterfall, really, but an ancient coral reef, now hardened into a mass of limestone, that lay exposed in low water. Early pioneers could pass this stretch only a few months out of the year, when there was water enough to fill three narrow passageways. In the steamboat era, workers had waded into the shallow water, carrying rods and sledgehammers, opening holes that they packed with gunpowder. A more lasting solution was provided when a private company, supported by federal funding, opened a canal in 1830. Its three lock chambers offered an alternate route past the falls. But now the steamboats working the river were so much bigger that many could not fit inside. Congress gave the canal's owner permission to expand the locks in 1860, but the project was slowed by war. In 1867 Humphreys sent an officer to Louisville to oversee the project's completion. Five years later the locks had tripled in size. Congress bought out the private stockholders in 1874, turning the lock into a federally owned project.

Ironically, all this bustle served little purpose. When Mark Twain had been a pilot, he'd had to steer around rival packets and coal barges, sometimes even floating productions of *Hamlet.* When he returned to the river a few decades after the war to research *Life on the Mississippi,* he found the presence of even a single fellow vessel an occasion of note. He remembered that a hundred boats had once clustered at the St. Louis waterfront, where throngs of workers dashed about. Now he found "half a dozen lifeless steamboats." The wharves lay empty for a mile. But in a line of strained logic that would mark the river for decades to come, the lack of steamboats became the very reason to build new river infrastructure.

The railroads had by now become a problem. As the steamboat lines went out of business, shipping rates began to rise, prompting widespread fury. What was to be done? Passing heavy regulations on railroads seemed untenable; creating a government-run, public rail service seemed even more un-American. But fixing up the rivers was deemed an acceptable way to tip the scales. Some boosters didn't seem to care whether the improvements put boats on the rivers at all. "If there were not a pound of freight carried upon the waterways, the benefits derived from their presence would be incalculable, standing as they do a menace by nature to all artificial competition," one congressman later proclaimed. "The waterways are the most powerful regulators of rates upon railways." The result was an agency kept busy, smashing through ecosystems, spending money, to little notable effect. Among fellow engineers, the army offices became a kind of backwater. "In the rapid professionalization of engineering after 1850," one historian suggests, "the influence of military engineers was negligible."

At least one project in this era stands out as a real accomplishment. Tellingly, it was not the work of the army.

Over its final miles, the Mississippi River fractures, splitting into several channels. As the water spreads, it slows. Mud settles. This process built the delta landscape but also filled it with perils. The weight of the shifting silt can create mudlumps, for example—volcanic explosions that burst upward, venting methane and river sludge from their peaks. Their presence may have helped discourage early explorers from approaching close enough to discover the river's mouth. But in the American era, a much more mundane sort of lump became a crisis: the mud settled into shoals in the passes at the river's mouths, blocking the passage of international cargo.

This was an old problem; even the first French colonists had dragged iron harrows through the riverbed, attempting to keep the passes open for oceangoing freight. The shoals were another major focus of Humphreys's study. As chief engineer, he'd embraced a high-tech approach: he commissioned a dredgeboat to clear the channel, outfitted with three engines and two screw-blade propellers, each twelve feet in diameter. These were meant to kick the mud up to the river's surface, where it might be carried away by the current. The vessel was named the *Essayons*, which is the Corps of Engineers' motto—French for "Let us try." The name was unfortunately apt, as it turned out. Describing the boat's disappointing performance, a writer in the *Times-Picayune* opined that "the most she can do is break her propeller, and steam up to the city for another." Eventually, Humphreys embraced a simpler solution, one that leaders in New Orleans had proposed: just bash through the river's natural levees to create a canal. Boats could then exit the river into the gulf forty miles upstream of the problematic passes.

By 1874, when a Missouri senator introduced legislation to set aside the funds, the plan seemed likely to proceed. Then just days later a civil engineer named James Buchanan Eads arrived in Washington. He proposed to revive a concept the military had tried and rejected decades earlier: build a set of jetties to concentrate the flow of water enough that it would blast through the shoals. Such structures, Eads said, could yield a channel deeper than any canal that the Corps of Engineers might dig and at lower cost. And

Eads offered the sweetest of financing deals: he'd pay for construction, taking no payment until the government could see that his idea had worked.

The American Society of Civil Engineers, launched in 1852, had by now helped the nation's new crop of engineers coalesce into an institutionalized profession. *Civil,* back then, meant *civilian,* and it was a label the engineers wore with pride. They liked to point out that unlike their army counterparts, their careers depended on actually succeeding in what they built; they could not depend on the security of military rank. Besides, the graduates of the new university programs thought their training was superior to the quick, two-year curriculum at West Point.

James Buchanan Eads was, in this context, an odd man out: while he had not gone to West Point, he had not gone to school at all. His family had arrived in St. Louis by steamboat in 1833, seeking a fortune. Eads, at thirteen, had to abandon his formal education to sell apples in the streets, which suggests how this mission fared. At some point, young Eads talked his way into a job with a merchant who noted the boy's spunk and intellect. The man granted Eads use of his personal library. Thus began a classic rags-to-riches story: one of the most accomplished engineers in history taught himself the discipline by flipping through books in his off hours after work.

A few years later, when the rest of the family moved on again, Eads made his first bold decision: he'd stick it out and find his way. St. Louis was a town of destiny, sitting atop the bluffs where two rivers meet. Seventy years after its founding, it offered a portrait of the nation's changing culture. Buckskinned hunters strolled past merchants in business suits. As a sixteen-year-old, Eads took a job as a mud clerk for the steamboat *Knickerbocker.* He sloshed along the unpaved waterfront, signing for the arrival and departure of cargo. When in 1839 the *Knickerbocker* struck a snag near Cairo and sank, losing its load of lead, Eads noted a problem to be solved. Later, he conjured a design for a steam-powered salvage boat that could lift such wrecks from the riverbottom. As a twenty-two-year-old, equipped with a few drawings and a commanding confidence, he convinced two boat builders to construct the thing, in exchange for a stake in the resulting company.

Even before the boat was finished, Eads won a contract to retrieve a few hundred tons of lead that had been lost in the rapids near Keokuk, Iowa. He rigged up a diving bell from a whiskey barrel, one that he figured would work in the river's strong current. When the diver he'd hired from the Great Lakes refused to wear the device, Eads donned it himself and had his crew swing him into the river. He sank to the bottom, where, fumbling in the darkness, he tied his cables around the lead pigs. Soon he had a booming business. As St. Louis became a gateway town for prospectors rushing west to California, he told his wife he'd already found his gold mine. He eventually walked the bed in hundreds of places. In one case, sixty-five feet below the river's surface near Cairo, as he tried to find footing, he passed right through a mass of mud "until I could feel, although standing erect, the sand rushing past my hands, driven by a current apparently as rapid as that at the surface."

Eventually, Eads moved on to bigger ventures. When the government closed the Office of Western River Improvements in 1856, he bought the fleet of snagboats and took funding from fifty insurance companies to keep the river clear. He retired the next year as a thirty-seven-year-old with a half-million-dollar fortune. Then the Civil War broke out, and Eads was lured back to work building ironclad gunboats for the Union Army.

After the war, Eads proposed a bold design for a bridge in St. Louis. It would be the first to cross the river downstream of the Missouri, whose waters turn the river wide and powerful. When completed, it would be the first large steel structure in the world. His design was influenced by his knowledge of just how chaotic the riverbottoms were, but was also so novel that it terrified many of Eads's fellow engineers. One of his own employees, an experienced bridge builder whom he'd hired largely to reassure investors, called the design "entirely unsafe and impracticable." Many others in the field agreed. The criticism partially rooted in their own failure to understand the mathematics, but it could not have helped that Eads was an outsider, an autodidact rather than a carefully trained graduate of the new schools.

Eads, as willful a man as Humphreys, quickly fired the underling who'd criticized the bridge. Perhaps this is just the way of the engineer. Eads liked to believe he was smart enough to ascertain every "phenomenon and eccentricity of the river," as he once declared in a speech. The river's "scouring and depositing action, its caving banks, the formation of the bars at its mouth, the effect of the waves and tides of the sea upon its currents and deposits"—all these were "controlled by laws as immutable as the Creator, and the engineer needs only to be assured that he does not ignore the existence of any of these laws, to feel positively certain of the result he aims at." To consider yourself a master of so much chaos is to place yourself above the realm of other men.

As his bridge went up, Eads decided his river expertise qualified him to weigh in on Humphreys's plan to dig a canal downstream of New Orleans. At a conference in St. Louis, in front of hundreds of congressmen, Eads dismissed the concept—and then afterward, on an excursion to the river's mouth organized by the Corps of Engineers, expounded on his theories about jetties. Perhaps he should have known he was about to start a war: this was not the sort of talk that the army's chief engineer would countenance.

Humphreys, as the head of the Corps, had the right to review any bridge spanning a navigable river, a policy initiated after the 1856 fiasco in Davenport. Eads's St. Louis bridge plan had passed early inspections, but now, with the structure nearly finished, Humphreys intervened. The Corps declared it a hazard to steamboats; eventually, the army engineers issued a report that described the bridge as a "monster . . . that must come down." But Eads knew his dance moves as well as Humphreys did, perhaps even better. He took a meeting with President Ulysses Grant. The two men walked the spans of the nearly finished bridge, then retired to Eads's office to warm up over brandy. Based on those discussions, Eads decided he could safely ignore the army's commands.

Then he struck back: Eads traveled to Washington to upset Humphreys's nearly settled plan to dredge the canal. Even if some of his engineering colleagues were wary of Eads, his proposal gave them an opportunity. The attacks on the military engineers reached a new pitch. One congressman suggested that the army wanted to build its canal partly to satisfy a few wealthy allies: the men who owned the land where the channel was to be dug would profit from the sale, and they had the army's ear. That whiff of corruption launched talk of a new system, one that would dilute the power of the army's officers by including civilian engineers in the decision making.

Whatever Eads's motivation, Congress was happy enough to eat his free lunch. So in the summer of 1875, crews arrived at the river's mouth and began to lay a series of wooden pilings into a line that would stretch more than two miles beyond the end of South Pass. Then the men—who, lacking any firm ground for quarters, slept stacked in bunks on barges—harvested willow from the delta's marshes. They loaded the trunks by the hundreds onto barges, to be shipped thirty miles south to the pass. There they were woven into "mattresses" that were secured to the pilings, forming the heart of the jetties' wall.

Despite the disparaging "memoranda" Humphreys tucked into his reports to Congress, and despite his effort to spook Eads's investors by leaking false and damaging data, Eads pushed forward. Even before the jetties were complete in 1879, large vessels were already steaming through the pass. "Captain Eads, with his jetties, has done a work at the mouth of the Mississippi which seemed clearly impossible," Mark Twain wrote a few years later, when he published *Life on the Mississippi*. The old steamboat pilot had always been skeptical of the engineers. In the wake of Eads's accomplishment, though, he did "not feel full confidence . . . to prophesy against like impossibilities."

And Eads was not yet finished with impossibilities. When invited to speak to Congress, he announced an even wilder dream: straighten the river's looping bends to speed the current, then create a line of dikes directly upon the riverbanks. Turn the entire river into one long jetty, in essence, cranking up its erosive power so it would dredge its own bed. To help sustain this rip-roaring river, Eads wanted to shutter the existing bayous in the delta—something that even Humphreys, despite his commitment to levees, had never endorsed.

Eads's larger point was that the Corps of Engineers should not command this river alone. Congress agreed and in 1879 created the Mississippi River Commission, a board that included three army engineers and three civilians, plus, as a kind of peacekeeper and tiebreaker, a representative from the U.S. Coast and Geodetic Survey. Congress gave these men an ambitious remit: to "correct, permanently locate, and deepen the channel and protect the banks of the Mississippi."

Disgusted at another loss to Eads, Humphreys retired from the Corps of Engineers two days after Congress passed the bill. One week later, as if to mark the completeness of Eads's victory, an army official confirmed that the jetties had met the terms of the contract. New Orleans, which had shriveled into a middling port, soon rose to become the nation's second largest. But Eads's own tenure on the commission lasted less than four years; he quit in 1883, once he realized he'd never convince his colleagues to pursue his grand impossibilities. So just a few years after their battles ended—at the moment when the nation's ambitions for its waterways reached higher than ever before—neither of the bull-headed engineers retained any say over what would happen to the Mississippi River.

When it came to the heart of the matter, Eads and Humphreys had always agreed: the river was not just knowable but tamable—a thing to be conquered. Now, without them, the work to tame and conquer the river continued. More money was spent on river improvements in the 1880s than in the previous six decades together. Hoping to deepen the Upper Mississippi to six feet, the Corps closed backchannels with walls of willow brush. (The logic was akin to Eads's: narrow the river's choices, and it would speed through the path that remained.) The engineers were at work on the river's first reservoir, too, meant as an experiment that might prove the viability of storing water so it could be gradually released, keeping the river at its desired depth. The dam, finished in 1884, sat on the edge of Lake Winnibigoshish, four hundred miles upstream of St. Paul. The land along the river here still belonged to the Ojibwe, who watched in horror as water swallowed cranberry bogs and hunting trails and fishing shoals and wild rice marshes. Ancient burial mounds were washed away; skulls and bones lay scattered along the new shoreline. Fish lay dead in the river. The camp built for the workers became the first white settlement near the headwaters; the Corps built roads and telephone wires to connect the site to the grid. The dam soon turned into a tourist site, beloved by sports fishers.

As for the Ohio, after years of debate and discussion, a tentative effort was underway to turn the whole river into slackwater: blocking its channel with dozens of dams might ensure the water was deep enough for navigation all year, the engineers hoped. This approach had already succeeded on small rivers throughout the Ohio Valley; now work was underway on an experiment near Pittsburgh, to see if such a big river could be converted into similar pools. Men labored fourteen hours a day slicing stone out of mountains, then laying the rock into the river to build the walls of a new lock. The Davis Island Lock and Dam, as the experiment was known, opened in 1885. Reporters estimated that fifty thousand people thronged the riverfront to enjoy the speeches and songs that accompanied the locks' dedication. They may have been disappointed: a broken pump kept the water from draining properly, and the first vessel to pass—a boatload of cabbages—had to wait until the next day.

As for the Mississippi River Commission, it did not take long for its supposed independence to evaporate. By law, its president had to be a military officer, and the agency was never given its own staff. The commission officially held jurisdiction over the whole of the Mississippi River, north to Lake Itasca, but its members endorsed the Corps's ongoing work along the upper reaches of the river and left them to the task. Even on the lower river, where the commission created four geographical districts, an army officer was placed in charge of each. By 1882, Twain was describing the river's new masters with a suitably blurry phrase: the "military engineers of the Commission."

Twain saw the work underway when he traveled the lower river in 1882: the commission was erecting new wing dams to send the current in new directions and new dikes to narrow the flow. For "unnumbered miles," Twain noted banks newly emptied of trees so they could be graded and protected in stone against erosion. Rows of piles held sodden shorelines intact. Even in the wake of Eads's success, this all seemed a bit ambitious to the old pilot. "One who knows the Mississippi will promptly aver," he wrote, "that ten thousand River Commissions, with the mines of the world at their back, cannot tame that lawless stream." But, then again, this was the Army Corps: *Essayons*.

Farmland in the alluvial valley (Bolivar County, Mississippi).

7

That Big Green Wall

The frontier disappears—and the floodplain turns tamed

When a young man named Will Dockery delved into the alluvial jungle in Mississippi in 1888, there were still few roads through the swamps, and fewer bridges. So when Dockery came upon a bayou, there was nothing to do but wade across. It took three days to travel just eighty miles, from Dockery's family farm near Memphis south to the frontier town of Cleveland, Mississippi—which consisted of little more than two stores, a saloon, and a two-story hotel, built alongside a newly installed rail depot. This town was deep in the floodplain, miles from the river, in what until recently had been considered impossible country. The streets were little more than a network of mud troughs.

That the floodplain remained so wild was a strange outcome, given the river's history. The Mississippi had been a magnet for decades, tugging the nation westward; New Orleans had exploded into a major city, and for a hundred miles upstream, all the way to Baton Rouge, the banks were lined with lavish plantation homes. But the alluvial valley, despite the richness of its soils, had proved a bit too difficult. "Human life along the Mississippi is indescribably insignificant," Frederick Law Olmsted had written back in the early 1850s, as navigation on the river hit its boom years. Compared to the few towns he saw in the alluvial valley, the steamboat Olmsted rode seemed to "rank a great metropolis." A decade later, in his survey, Andrew Atkinson Humphreys referred to the Yazoo Basin as "that great Swamp." While the local population more than doubled between 1850 and 1860—reaching 50,000, most of whom were enslaved—just 10 percent of the forest had been cleared. Across the river, in Arkansas, the trees stood even thicker. The trouble, of course, was that the Mississippi River refused to remain in its bed.

The war brought new challenges. The Union Army smashed through the levees as a part of an ill-fated quest to sneak through the bayous and lay siege to Vicksburg. The trek through the swamps proved so miserable that simply arriving back on the big river after six weeks was considered victory and celebrated with a musket salute. After the war, in 1866, Humphreys's trusted lieutenant Henry Abbot traveled the Mississippi, marking crevasses on his map in red ink. "The river is looking very badly," he wrote. In Louisiana alone, more than a hundred miles of levees had been destroyed. The cotton fields in the Yazoo Basin lay abandoned; fresh cane and willow trees reached twenty feet tall. Bears smashed through the few remaining cornfields. Alligators stole pigs.

Abbot marshaled what labor he could find—occupying Union soldiers, Irish immigrants, prison gangs—to create slapdash replacements for the broken levee. But his orders from the War Department made clear that, since flood protection was the duty of local governments, he was not to conduct systematic repairs. So more than three hundred years after Hernando de Soto crossed the Mississippi and found in its swamps so many wealthy chiefdoms—nearly two hundred years after La Salle claimed the watershed as a part of a white empire, two generations after the Louisiana Purchase, and one generation since the treaties that had removed the Choctaw and the Chickasaw—most of the river's alluvial valley remained empty of people.

To Will Dockery, that made this place enticing. A new fortune might be made. His father was one of the men who had given up the exhausted soils beyond the Appalachian Mountains in the 1850s and traveled west; the family had settled in the rolling hills just north of the Yazoo Basin. The war must have been a strain on the Dockerys, but after a few years in college, Will was able to take a job as a clerk in his uncle's store in Memphis. And there was money enough in 1888 to send Will south to set up a new branch of the family operation.

The timing was good: the floodplain was just beginning to boom. Within two years of setting up his small storefront in Cleveland, Dockery had made enough money to lease some eight hundred acres of land and fulfill the dream of that place at that time. He put a crop of cotton in the ground.

"The country was covered with blue cane fifteen to twenty feet high," Dockery later remembered, "and the land was rich as cream." Despite the potential, the ancient trees were so hard to free that even people who had bought land on the cheap—five dollars an acre—would quit after a year, willing to trade it even cheaper. Forty acres could go for as little as a single cow or a Winchester rifle. Dockery, though, was not going to quit.

Back in 1801, at his inauguration, Thomas Jefferson had sung the praises of his "chosen country" and noted how spacious the place was. It had "room enough for our descendants to the thousandth and thousandth generation," he suggested. He rather missed the mark, as it turned out. It took less than two generations for the railroads to arrive and the United States to spill out of the Mississippi watershed; before a hundred years passed, so many white people had moved so far west that the U.S. Census Bureau could no longer discern a line that marked the frontier.

With so much landscape swallowed, the little bits of wildness left took on new value. John Muir's dispatches from his treks up the Sierra Mountains, now considered classics of nature writing, were first published in the 1870s. Less well remembered are the wilderness pilgrims who ventured onto the Mississippi in this era. After the travel writer Nathaniel Bishop hiked a thousand miles across South America and slipped down the Atlantic coast in a paper canoe, he decided the western waters offered a suitable follow-up challenge. He departed Pittsburgh in 1875 in what he called a "sneak-box," a small duck-hunting boat with a shallow draft and a rear paddlewheel. Over four months he descended to the gulf, then continued on to Florida.

Six years later another travel writer followed a similar route canoeless; he strapped himself in a rubber suit that featured five air chambers and lay on his back, using his body as a kayak. At least two more Mississippi River travelogues appeared over the next few years, all following the path that the flatboatmen had forged nearly a century earlier. The farm boys were gone now, another victim of the railroad's rise in the years after the Civil War, so Bishop and his fellow voyagers were reviving a lost American adventure. The western waters weren't a primeval place; on the Ohio River, the writers noted derricks rising from the surrounding mountain slopes and sheets of oil staining the river's surface. But the idea of wilderness was new enough that it could accommodate such sights. Bishop sounds at times like a lesser version of Muir: in his sneak-box, he claimed to be throwing aside "the chains of every day of life" to become "master of his own time, and FREE." The river's wildness had become something different now: not an obstacle to conquer but a pleasing counterpoint to a world that had, to borrow a phrase from Huck Finn, turned "so cramped up and sivilized."

The floodplains represented something different than the river itself. Yes, these uncleared forests were relict slices of wildness, and you had to be adventurous to trek into their swamps, but the point was not a diversion. The point was to claim land. Here, then, was one more task for the great engineers. After the war, they began to extend the levees northward, beyond the lower river.

Illinois's largest levee district was chartered in 1871, taking responsibility over fifty-four miles of land north of St. Louis—hundreds of thousands of acres of potential farmland. The engineer who led the district, Elmer Corthell, noted that the swamps here had long been sickly, marked by "desolate farms and despondent individuals." Five years later, the *St. Louis Times* described the change: "A Malarial Swamp Made to Bloom Like a Rose," the headline declared. (Corthell by then had shipped south to join Eads's team, constructing the jetties.) But along the Lower Mississippi, a much bigger river, the task of stopping floods had always been harder.

In 1874 Congress commissioned another study of the southern flood problem. Leaning heavily on Andrew Humphreys's conclusions, the committee of engineers concluded that only a centralized project could stop the river's assault. Building new levees, the engineers suggested, could be a way to unite the nation: wash away the memory of the recent divide by waging a new kind of war, a war against the river—that "terrible enemy," their report called it. The alluvial lands should be divided into six districts, each reporting to some as-yet-to-be-identified federal authority. This had been the one point of agreement among the warring engineers for decades: Humphreys and Charles Ellet both had figured that to beat the floods, the federal government would have to get involved.

Northern legislators refused to budge. Why should everyone pay so as to redeem the rebellious South? With the southern economy in shambles and no hope on the horizon, some old planters began to sell their uncleared land to stave off bankruptcy. Many who didn't found their land seized for unpaid taxes. By 1878, more than half of the Yazoo Basin belonged to the state of Mississippi.

This offered its own sort of opportunity. Land prices were low enough—a few hundred acres might sell for a thousand dollars—that even Black families, only recently freed from slavery, could find a foothold. Decades later, at the close of the nineteenth century, as many as three-quarters of the landowners in some of Mississippi's riverside counties were Black. They were one part of an ethnic frontier hodgepodge: Chinese immigrants found a niche opening grocery stores that catered to the Black residents, a clientele whom the white businesses hated to serve. Lebanese merchants traveled the soggy roads, peddling wares in the newly established frontier towns.

The railroad magnates saw opportunity here, too, and bought up huge tracts. One company alone acquired two million acres in the Yazoo Basin in 1881. As they laid their tracks, these railroad companies reshaped the floodplain's geography. Before the war, the few plantations in the alluvial valley had faced the local rivers, which, after all, were the only paths through the wilderness. Commercial life was bound by the shape of the watershed. But the names of the new rail lines, often rendered in acronyms, pointed in new directions: the LNO&T, the first to arrive, meant the Louisville, New Orleans, and Texas Railroad Company. The T&SL was the Texas and St. Louis. Suddenly, soggy forests were linked to major cities. The floodplain's interior, long considered inhospitable backcountry, was open for settlement.

Lumber towns appeared wherever the tracks ran. They could be rustic places. "We could hear wolves howling, see bear tracks, and hear raccoons fighting over the cans we threw out," one foreman's wife remembered of a turn-of-the-century lumber camp. "For months and months I didn't see a single woman." She didn't mind the bears, she said, but the wolves were a problem, as were the panthers: She just never could get used to their nighttime screams.

Some of these towns lasted just a few years—until the trees were gone—but others blossomed into civilized villages. As they grew, and as the steamboats sputtered, the few towns that had once stood along the big river lost their purpose. In 1880 Australia, a small village in Bolivar County, Mississippi, still included a church, a doctor's office, and two stores. Four years later and fifteen miles inland, along a bayou that was too small to accommodate the steamboats, an LNO&T construction crew erected a shack as their depot. Then the men built a house as their quarters. A post office and a general store soon followed. Today Cleveland is the county seat, while Australia is a willow grove on the wrong side of the levee, inaccessible by public roadway, a ghostly name on the map.

Will Dockery arrived in Cleveland in 1888 on the first crest of a new tide of settlement that would last half a century. The timber camps were one draw: every tree that came down meant more potential land for farmers. The railroads helped, too; by 1906, 90 percent of the land in the Yazoo Basin lay within five miles of the tracks. The trains delivered flocks of eager settlers, seeking fortunes. Perhaps the biggest help, though, came from the Mississippi River Commission, which gave the new settlers hope that the floods might finally be stopped.

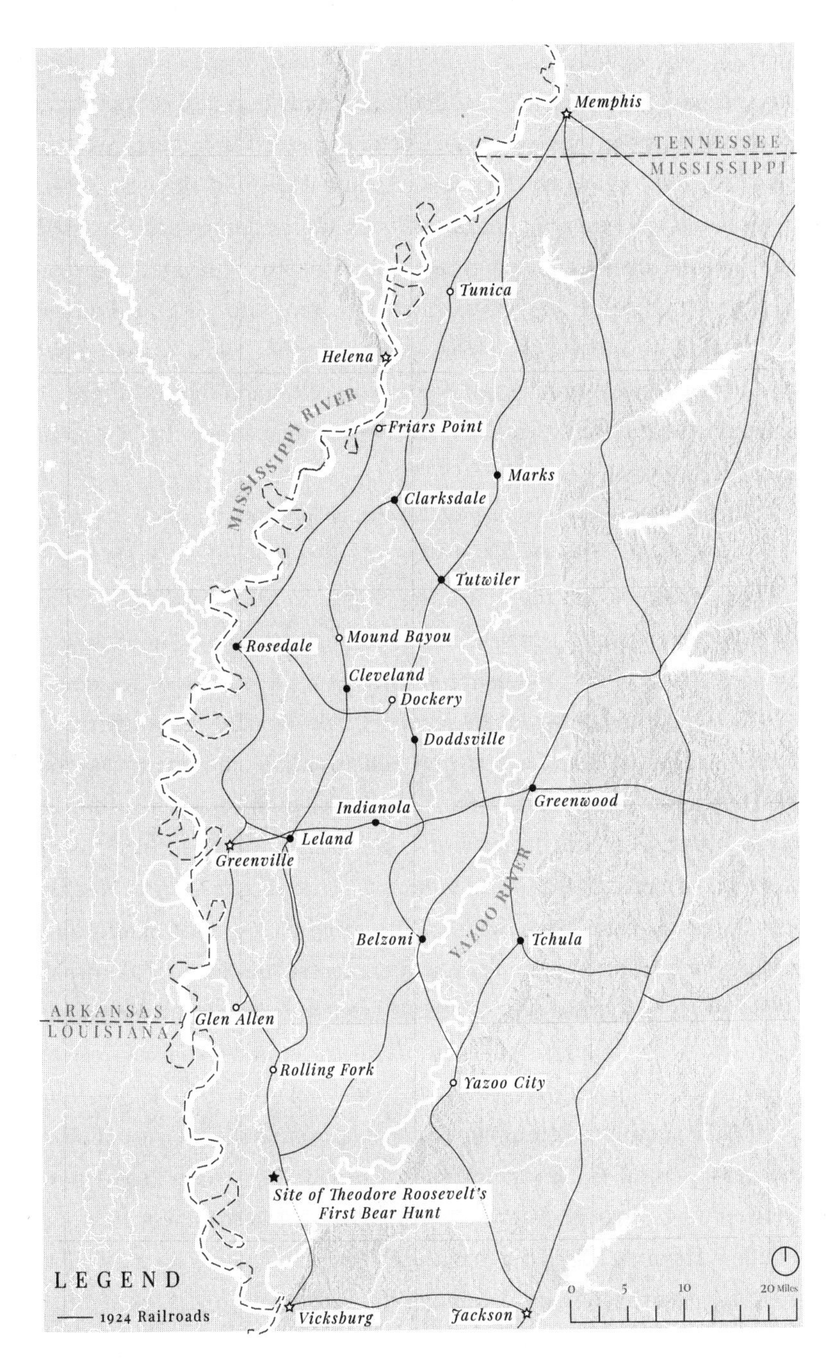

Railroads, levees, and drainage canals helped turn the interior of the Yazoo Basin into an expanse of cotton farms.

Back in the 1870s, when James Buchanan Eads began to expound upon his river impossibilities, Southern interests noticed an opportunity. Eads wasn't focused on levees; the dikes he wanted to build along the Mississippi's banks were meant not to stop floods but to compress the low-water currents so they would deepen the river. This was a navigation project, in other words, which did fit within the federal purview. But Eads was proposing a wall along the river, something that felt a half-step away from a flood-protection levee. Southern politicians figured there was something useful here, which was why they pushed for the Mississippi River Commission.

It's worth pausing to describe Southern politicians of this era. For a brief a moment after the Civil War, Blacks acquired not just land but real political power in the South, serving in Mississippi as sheriffs and tax collectors, even as senator and lieutenant governor. Then came the backlash. A few years of violence culminated in the Mississippi election of 1875, which, in the words of scholar Clyde Woods, "began with cannon fire aimed at Black voters and ended with numerous massacres." The Democrats—the party then of the white planters—cemented their power. Ulysses S. Grant declared that Mississippi was now ruled "by officials chosen through fraud and violence." The old guard, in power once more, began to piece back together the patchwork of vast plantations—and expand, too, into newly cleared lands across the river in Arkansas.

Two years later, after another heated election, Republican Rutherford Hayes was seated as president only after he promised concessions to this Southern bloc—including infrastructure improvements. Hayes made good on the promise, signing the law that convened the Mississippi River Commission in 1879. The commission was widely seen as a Trojan horse, an attempt to turn privately owned swamps into farmable land. And President Hayes, still hoping to placate Southern Democrats, stocked the commission with engineers happy for the task.

At first, the commission declined to ruffle feathers, focusing on constraining the low-water channel. But a bad flood in 1882 prompted the commissioners to ask for, and receive, permission to repair the levees.

This was still nominally a navigation project: the claim was that leaving a gap in the wall might lead to damaging ripple effects down in the navigation channel. But this episode marked a turning point. Every subsequent Mississippi River Commission budget provided at least some funding for levee projects, and by the 1890s—as Will Dockery, backed with credit from a wealthy uncle, expanded his cotton farm into a sprawling plantation—the federal government was outspending states and local districts on levee construction. The commission's 1895 report openly bragged about how much land the engineers had been able to protect from the floods.

The dike- and dam-building projects that the commission had first pursued, meanwhile, had failed to do much in the way of "correcting" the riverbanks. The military engineers had tried, but they'd failed. Now the commission suggested they give up on such work: the Corps might as well just send out dredgeboats to keep the channel open, it suggested. Not that the commissioners intended to close their office. They just thought their new focus should be on helping the local districts build more and more levees. Congress agreed—and so the great wall-building project commenced.

Decades later, when the ethnomusicologist Alan Lomax visited the land along the Lower Mississippi on a quest to discover unrecorded blues songs, he became obsessed with the levee. "Over the years this great wall of earth has loomed ever larger in my imagination," he wrote. On one trip, Lomax picked up a hitchhiker, an old Irishman who had worked on the levee as a day laborer. Lomax fished for stories. The man's wages, he learned, consisted of "a few dollars and so many jiggers of whiskey." The Irishman was eventually promoted to foreman, not because of his talents but because he was outworked by his Black peers; by the time the Mississippi River Commission began building levees, the companies they contracted had decided the Black workers were far more proficient at whipping the mules—"skinning" them, to use the phrase of the day. The animals dragged "wheelers," a kind of bulldozer, as well as carts filled with dirt that was dug from pits and mounded along the river. They pushed their mules with "wild work calls" that Lomax identified as a key ancestor of the blues.

Each worker had his own tent, though some shared their space with women, who were welcome in the freewheeling camps. The floors were dirt; piles of soil mounded along the edges of the tents kept rain from seeping inside. Arranged along gridded, muddy streets, the camps could house twenty-five or fifty or even hundreds of workers. (One early Louisiana project in the 1880s required the labor of 700 mules and nearly 4,000 men, including several hundred convicts.) Toilets were a too-expensive luxury; buckets of waste must have been emptied into the teeming swamps. "You could smell those camps a mile away," an engineer told Lomax. "There was a buzzard on every fencepost." When the river rose, the camps would look like islands, accessible only by boat. They featured a commissary that sold cocaine (among other staples) and where, on days too wet to work, the men gathered to play craps. But on days with better weather, the muleskinners rose before dawn. Work could begin at three in the morning and proceed with few breaks for twelve hours. It took months to progress just a hundred yards.

Lomax described the levee camps as "the last American frontier, even more lawless than the Far West in its palmiest days." They doubled as hideouts for fugitives, because the foremen kept the law out. (When a humane officer complained about the poor treatment of mules—not the muleskinners, of course, just the mules—the contractor who ran the camp shot him. A payment of $50,000 untangled the contractor from the legal mess.) One blues artist who worked in the camps called them the "toughest places I've ever seen." The stories suggest that everyone carried a gun, and a white engineer in 1904 noted that to see a foreman shoot and kill a muleskinner was common. The work wouldn't stop; the corpse was buried in the night, "and that's all there is to it," he declared. Black laborers were seen as disposable, tools to be used in the fight against nature. That metaphor could become literal: during a flood in 1912, an engineer in Greenville, Mississippi, ordered Black prisoners to lie down atop the levee, using their bodies to hold back the rising water until sandbags could arrive.

To help corral labor, contractors delayed payday, sometimes by months, forcing workers to shop at the commissary, where prices were two to three times higher than in town. Even workers who decided to live outside the camps could be charged for room and board. In the end, the muleskinners practically worked for free. "I can get all the labor I want for whatever I care to pay," one contractor told an undercover investigator.

The muleskinners thought of themselves as "rough mens," as one put it, using the local grammar. The camps were brutal, but they were better, perhaps, than working in the fields. Workers could claim a small and broken autonomy: the foremen let them fight and drink and gamble and whore, so long as they did their work. (Often the foremen themselves sponsored these after-hours activities so they could claw back what little salary the workers were actually paid.) Some of the exploits are exaggerated—it's hard to know what is blues-song romance and what is the levee-camp truth. Some historians believe the extent of shootings has been overstated, but it's clear that whippings were a frequent punishment.

As these men piled up the levee, the river beyond seemed to grow ever more romantic in the American mind. In 1885, a decade into the Mississippi travel-writing boom, Mark Twain released *The Adventures of Huckleberry Finn*, which soon became a sensation.

The novel is set, as a prefatory note indicates, "forty to fifty years ago"—back in the bygone days before the nation had swallowed its full western expanse. But the book offers the kind of nature worship more common in Twain's own era. The writer made the river into an American idyll: with his companion Jim, Huck floated the river by night; then, as predawn blackness paled into blue, he tucked the raft into the forests of cottonwood and willow. "Then we set out the lines," Huck says. "Next we slid into the river and had a swim, so as to freshen up and cool off; then we set down on the sandy bottom where the water was about knee deep, and watched the daylight come." In Twain's wake, even the desperate shantyboaters were cast in a rosy light. A reporter for *Scribner's* visited in 1900 with a clan who, he said, liked to think of themselves as "the river people." To claim your place among this tribe, "you must be lulled to sleep at night, in your own floating home, by the gentle lapping of the water, all the year round," he wrote. "You must be capable of living mainly on fish which you have taken from the river's tawny depths. You must cook your food over driftwood fire."

The swamps in the floodplain had found their admirers, too. Wealthy men founded "sporting" organizations by the hundreds in this era, groups that sought out what wildlands remained so they could enjoy the old-fashioned thrill of the chase. The most famous, the Boone and Crockett Club, was launched in 1887 by Theodore Roosevelt. (It's named for two of his favorite men of the western rivers: Daniel Boone and Davy Crockett, who had earned fame amid the watershed's rough-and-tumble frontier years.) Now the same rail lines that carried lumber out of the forests delivered businessmen in. They spent their days shooting birds and their evenings smoking cigars and drinking whiskey from the comfort of a club lodge. At official meetings for one swampland club in Arkansas, the hunters wore headdresses, playacting at the lives of the people who had once been considered the enemy.

To their credit, these hunters knew their prey needed habitat. As president, Roosevelt protected more than 230 million acres of public land—national forests and national parks and national game refuges, land that would be set aside from the rush of development. As the system grew, it eventually encompassed hundreds of miles of the Mississippi: in 1924 Congress turned the strip of marsh and forest between Wabasha, Minnesota, and Rock Island into a massive wildlife refuge. This declaration offered a major exception to the trend of the era: no levee districts would be formed here; the floodplain would stay connected to the river. Roosevelt himself was more focused on the Lower Mississippi. The expanse he protected included several barrier islands at the edge of the river's delta, which were designated federal bird reserves, and he took several personal trips south, into the alluvial valley. On one of these voyages, he dressed in a costume of fringed buckskin. He was seeking the ultimate floodplain adventure: he wanted to kill a bear.

His first attempt, in 1902, was famously unsuccessful. Roosevelt declined to shoot the emaciated animal that his guide had tied up on his behalf—thereby inspiring toymakers to invent the Teddy bear. (What is less often recounted is how the sad old bear was ultimately dispatched. After a knife to the ribs, the corpse was slung atop a horse and carried back to camp, to be butchered and served at meals.) Roosevelt was more successful a few years later, in Louisiana, in a hunt he recounted in an essay published in *Scribner's*.

The document captures the schizophrenia of the era: Roosevelt begins by praising the rich soil of the floodplain and noting, with seeming enthusiasm, that once the levees were strengthened, the entire floodplain could go into cultivation. He details the mechanics of land clearing, then praises the beauty of the sections of land that had been cleared already—the rows of cotton, spread like snow across the landscape, broken up by small homes for the tenants. Then he turns romantic. "Beyond the end of cultivation towers the great forest," he writes, as he shifts into a backwoods adventure that culminates in his attack on a cornered bear. "I knew my bullet would go true; and, sure enough, at the crack of the rifle the bear stumbled and fell forward." She was not killed enough for Roosevelt's taste, since in her last few minutes she might strike one of his expensive hounds. So he "fired again, breaking the spine at the root of the neck; and down went the bear stark dead, slain in the canebrake in true hunter fashion."

The essay offers a reminder that there were always two sides to the frontier, and that Roosevelt wanted both. The levees and drainage canals would open new lands for development, continuing the great American economic bonanza; the refuges, meanwhile, would preserve a few slivers of forest, just enough space to retain the thrilling violence of frontier living. Roosevelt saw such brutality as an essential component of the American character: "A race of peaceful, unwarlike farmers would have been helpless before such foes as the red Indians," he declared in one of his several books of revisionist frontier history. There was a widespread worry in this era that without more land to seize, more battles to fight, Americans would turn soft. Less than pristine nature or functioning ecosystems, then, Roosevelt seemed to be seeking a way out—a way backward, a way to pretend that he was as hearty and independent as the half-horse, half-alligator rivermen he so admired.

The Mississippi itself, that great path for navigators, brought out the other side of Roosevelt's American schizophrenia. Perhaps because he descended from a long line of inventors—his great-grand-uncle, Nicholas Roosevelt, had been hired by Robert Fulton to design and build the *New Orleans*—he seemed unswayed by idyllic visions of the river. The channel, he believed, deserved a thorough reworking.

As president, Roosevelt convinced Congress to launch yet another agency, the Inland Waterways Commission, that he hoped would end the haphazard process of river improvement—the one-off pork barrel projects—and deliver to the western waters the kind of engineering triumphs that marked the era. He wanted a coordinated system, stretching from the headwaters to the mouth of the Mississippi River: levees to protect the floodplain; reservoirs to hold water for farms; dams to generate electric power; and restored forests to anchor the soils in rapidly eroding watersheds. Roosevelt's bear-hunting trip in Louisiana was the last stop on a steamboat tour, an official Inland Waterways Commission inspection of the soon-to-be improved rivers.

As it turned out, though, this new commission never did more than study the rivers. Congressmen preferred not to hand the reins over to some new agency; better to keep their old alliance with the Corps of Engineers, tossing it the projects they liked. And so the pork continued.

The plan for a full slackwater Ohio River was approved in 1910: fifty-four dams were to make the full thousand miles nine feet deep. The ambition was undeniable, though the expense soon seemed questionable. By 1916 a conglomerate of Pittsburgh coal shippers decided to stop sending product downriver to New Orleans, worried about the state of the southern river, where the Mississippi River Commission was more focused on building levees than on keeping the channel clear. Traffic on the Ohio immediately dropped by half. On the Missouri, a recent report had found that the federal government was spending more than a million dollars annually, mostly to install woven mats of willow trees on the banks, an attempt to stop erosion and tame the meandering channel. There was so little traffic that the navigation program saved shippers just $10,000 each year. On the Upper Mississippi, meanwhile, some years saw no through traffic at all. An editorial in 1926 in an engineering trade journal complained about the "psychopathic enthusiasts" fighting for further improvements.

Even the nomads seemed to have stopped their seasonal drifting—grown tired, perhaps, of smashing their houseboats against the wing dams, or suffering the claustrophobic experience of being shuttered into one of the concrete locks on the Ohio. The travelogues of the 1920s describe colonies on the edges of river towns, where the anchored shantyboats looked as if they had not been river-worthy in years. On a 1926 voyage, the only fellow wanderers that adventurer Harold Speakman encountered were two college boys, just graduated, who were aspiring Huck Finns: they figured they'd drift to New Orleans in their own messy houseboat, tie up to the bank, and launch a life as writers. To the north, on the upper river, passenger boats embarked after dark and traveled just far enough out of a city that the lights were no longer visible. Then, without any announcement, the band would stop its music; the lights and the engine were cut. "You could hear the watery stillness," Wallace Stegner wrote, describing the effect. "You were afloat on the dark river in the middle of the dark continent, in touch for a brief theatrical moment with the force and mystery of that mighty flood." But it was just a brief moment. Then the lights came. Real life resumed.

The levees, meanwhile, kept growing. They had to, to match the rising floods. The river reached record heights in 1912, ripping through the southern levees in fifteen places. The next year another flood roared down the Mississippi and burst through the spots that had just been repaired. In 1916 the river flooded again, breaking the four-year-old records. Even with assistance from the Mississippi River Commission, local levee boards could no longer afford the cost of repairs and expansions. So Congress dropped the pretense and passed a new law: the federal government was authorized not just to assist on levees but to take full charge of levee building as far north as Rock Island, Illinois. This special rule applied only to the Mississippi River. The rationale was just what Southern planters had suggested decades earlier: this was a national river, a national menace, that could be tamed only by national force.

To strengthen its wall, the Mississippi River Commission began to close the gaps in the levees where distributaries emerged. By the 1920s, then, besides openings that allowed the tributaries to deliver their water into the Mississippi, plus one outlet for the Atchafalaya River—a major distributary that, after extensive study, had been deemed too big to close—the wall stood unbroken, all along the southern river. Ninety percent of the floodplain had been cut away. "We levee contractors created a billion dollars' worth of land and property," one of the white bosses told Alan Lomax. "That's what we did down here in the state of Mississippi." The swamps had been tamed, turned into money, "and that big green wall protects that wealth."

So now, a hundred years after it first emerged, the cotton kingdom expanded to conquer a new province: the deep forests of the backswamp across the alluvial valley. A few small-time farmers remained, especially to the west of the river; the railroads and timber companies sold some cutover land in forty-acre parcels to Corn Belt farmers seeking lusher soils. But most of the newly clear ground went to wealthy speculators. And in the Yazoo Basin, where the swamps and forests were being cleared the fastest, the new tracts became attached to preexisting farms. By 1910, the U.S. Census of Agriculture noted that the plantation system was "probably more firmly fixed" in the Yazoo Basin than anywhere else in the country.

Plantation system is code for sharecropping, an approach first developed in the desperate years after the Civil War, when even men who kept hold of plenty of land had little cash to pay workers. Instead they offered "shares," small tracts of ten to forty acres, to tenants. The profit from the crop was split, though the bulk went to the landowners, who also tended to overcharge on goods from the plantation store. Some charged tenants fifty cents per acre for "insurance"—which was really insurance for the planter, since it funded an overseer who checked the tenants' work. Some tenants were white; many were Black families who moved into the floodplain in this era, hoping the sharecropping system might be their ticket to stability and wealth. The system did offer a semblance of independence: a tenant got to work his own plot of land, rather than on a wage-labor gang, which felt like a disconcerting echo of slavery. But it was hardly a route to prosperity.

Most of the Black landowners who had got their hands on plots in frontier days struggled to hold on. Small farmers suffered when cotton prices fell, as they could no longer cover their costs and had to take out loans. Then they suffered when cotton prices rose again, because the price for land jumped out of their reach. And the local banks were never particularly generous to Black borrowers. The big white planters, meanwhile, did not deign to deal with local banks, preferring the better interest rates offered in New York.

The new plantations were served by drainage districts that were appearing up and down the river to dig canals that would move the water off the newly protected land. The Little River Drainage District, which oversees a half-million acres in the Missouri Bootheel, may be the largest drainage project ever undertaken in the world. After the district was incorporated in late 1907, survey teams spent months charting the topography, shlepping from swamp camp to swamp camp in east-west lines, trudging through land where the water often lay three feet deep. Then what trees remained were torn free, to be stacked and burned. Dipper dredges were floated up the local bayous on barges. The crews worked three eight-hour shifts, around the clock, and slept in houseboats. Glossy pamphlets extolled the results of such projects. "The road to wealth and independence for the young man leads through the Alluvial Empire," proclaimed a 1920 essay that appeared alongside photos of modern homes and smoothly paved roads in the blossoming new landscape of the Yazoo Basin. As farmers plowed these new fields, they assembled the potsherds and arrowheads they found into private collections that could grow quite massive—a visible reminder that this place had been something else in an earlier era. But the earthworks themselves were being erased. By 1911, archaeologists had already noted that the extent of cultivation was so thoroughly destroying the old mounds that professional research was becoming "all but useless."

Today it's easy to project an old-world mystique onto the agricultural landscape in the river's floodplain: it looks like a perfect Arcadia, the final accomplishment of Jefferson's dreams. It's true enough that into the 1920s, the tenants' tools—mules and plows—remained ancient. But the farms themselves were more akin to modern corporations, where a central office tallied inputs and outputs.

New technology was coming anyway, as if to underscore what the business had become. Tractors began to take off in the early 1920s, though the new mechanical harvesters struggled to pull cotton bolls away from the leaves. So defoliants—calcium cyanamide, especially—were sprayed out of airplanes, along with pesticides meant to kill pests. The first commercial crop-dusting company was launched in Georgia in 1925 and quickly relocated to Monroe, Louisiana, in the heart of the alluvial cotton country. Later, it would become known as Delta Airlines, in a nod to the place where the company had found its first foothold.

Will Dockery had by then assembled a farm that spanned tens of thousands of acres along the Sunflower River, just east of Cleveland. The operation exemplified the plantation system, wherein whatever independence the tenants held was subservient to Dockery's feudal powers; it was like a town all its own, featuring a school, several churches, a dairy, and a train station with a full-time agent. Dockery Plantation even had its own currency for the commissary. Dockery owned another plantation, too, across the river in Arkansas, and chose to spend as much time in Memphis as possible to avoid the unpleasant alluvial life "among the bull frogs and mosquitoes," as he complained in a letter. The brash young city had become the Yazoo Basin's outlying capital, home to bankers and traders making a killing off the despoilation of the lands to the south. The planters would come up and gather in the lobby of the Peabody Hotel, a short stroll from the juke joints on Beale. These men took pride in running debt—a signal of their future gains. Land could be turned to money "so fast now," as William Faulkner put it, "that the problem was to get rid of it before it whelmed you into strangulation."

Cotton was particularly suitable to absentee landlords like Dockery: unlike corn or cattle, with cotton you knew your tenants weren't going to eat the crop. They couldn't even sell it until it had passed through the plantation gin. Local governments, meanwhile, would often accede to the authority of the men who ran these high-tech kingdoms. On one of his trips hunting for blues songs, Alan Lomax learned that to drive on a public road that bisected a plantation in Coahoma County, Mississippi, he needed to obtain a pass from the business's manager.

Here was a second form of control, a program of social engineering to match the levees and canals. Laws forbade tenants to sign new contracts without notifying their landlords; laws gave landlords near-complete control over how the cotton was marketed and sold. Curfews limited when Black residents could circulate. Regulations in some places forbade Black drivers to pass whites on the highway. Voting by Blacks was all but out of the question; attempts to organize were quashed. The worst came in 1919, when tenants and sharecroppers in Elaine, Arkansas, assembled a union. A group of local officials raided one of their meetings, and gunfire erupted from both sides. The next morning white militias descended on Elaine from across the region. They hunted down Black tenants, who fled into the canebrakes. Estimates of the death toll vary but reach into the hundreds. Afterward a dozen sharecroppers were sentenced to death; fifty or more were sent to prison. They'd been deemed guilty of murder, since a handful of white people had died in the violence.

Across the river, at Dockery's fiefdom, Saturday was a big day: thousands of laborers would gather at the commissary to wait in line for their pay. Usually, some musician would set up nearby, plucking out a few chords on the steel guitar—a shimmering, mournful sound that winnowed through the hot and heavy air. This first show was free, an enticement, a preview of the party to come. After dark the crowd of tenants and laborers drifted across the river, into a neighborhood of clustered shacks. One, known as the "frolicking house," would be cleared of furniture and stocked with mirrors and oil lamps. There was still no electricity at Dockery, so the frolicking house would flicker brightly above the darkened fields, calling to the masses. Inside, someone would sing and play guitar. The blues may have descended from levee camp hollers, but it was on the floodplain plantations where this music took its truest form. Dockery Plantation is famous today because it, like the levee camps, is considered one of the places where the blues was born.

Alan Lomax called the levee "the principal human response to the titanic power of the great river." That phrase suggests that such a wall was the only possible human response. Lomax seems to have forgotten that for most of a century, politicians had viewed the construction of levees as a waste of government funds. Our vision of this era can be skewed by the fact that most of the written record consists of notes and letters left behind by the planters like Dockery. They provide a picture of a lavish life in big homes, with dances and dinner parties where, despite Prohibition, whiskey flowed freely. This is a rather narrow glimpse of alluvial life in this era. The vast majority of the floodplain's inhabitants owned no land; they lived not in mansions but in small cabins. Their voices are harder to hear because, as Lomax noted, it was dangerous for Black residents to talk, or even sing, about how they were treated.

Still, the songs give clues of how people viewed this strange new alluvial empire, the endless expanse of cotton that had been assembled with taxpayer money and Black hands. "Every day seem like murder here," Charley Patton, the most influential bluesman at Dockery, grumbled in one of his songs. "I'm gonna leave tomorrow, I know you don't bid my care."

This was a common theme of the blues: wanting to leave, as fast as possible, as soon as you possibly can. Even today the colloquial expression among the region's Black residents suggests that your house is not the place where you "live," just the place where you "stay."

The ruins of the Mississippi Basin Model (Hinds County, Mississippi).

8

The Great Flood

A major disaster and the birth to a new kind of American river

Storms raged all across the heartland in August 1926—in Oklahoma and Kansas, in Indiana and Ohio and Illinois. The rain kept up for days, then weeks, so long that every hour felt like a dim, false dusk. So much rain fell that the soil could no longer absorb it, and it simply rolled across the earth's surface, joining the rivers and streams. The Floyd, the Sioux, the Neosho, Dry Creek: these rivers grew and surged until homes and farm fields were surrounded. Crops, unharvested, turned unharvestable. So it went for months, the water doing what water does—pressing downhill, seeking its exit. Throughout late winter and early spring in 1927, the Mississippi rose and fell, then rose again—then rose once more.

In March, buildings in the Smoky Mountains collapsed under the weight of record snow. The Tennessee River, roaring into a flood of its own, claimed a railroad bridge. By early April, a million acres across the Mississippi watershed were underwater, mostly in its upper reaches. Pittsburgh and Cincinnati were swamped. Fourteen people were swept to their deaths in Oklahoma. Some fifty thousand refugees had been cast from their homes. At Cairo, Illinois, the Mississippi hit the highest level ever recorded. In places, the river stretched three miles wide, from levee to levee. And the rain did not let up.

Most people then thought the levee-building era was coming to its end. The Mississippi River Commission suggested as much in its annual report in 1926, amid the gathering storms. The companies that had received federal paychecks and put Black men to work building the big green wall were preparing to shutter. The levees had reached the required grade; the work to come would simply be keeping the wall maintained. Now that hope seemed delusive. As the spring turned cold and the ground froze as far south as Mississippi, thousands of men were camped at the foot of the levee, piling sandbags day and night, in a desperate quest to hold the water. They lived in makeshift camps strung with electric lights and telephone wires. Their work was ceaseless. When a man fell into the current, no one bothered to pause.

Water can rage. It can also creep and burrow. Day after day, week after week, month after month, the water worried its way through the soils in the levees. It erupted on the far side into sand boils, little volcanoes of mud that grew ever wider as the water spilled out. In places, the levees shook like jelly. A footfall could open a muddy rift.

On March 29 a small crevasse opened on a backwater levee in Arkansas. April 8: another small crevasse, in Alexander County, Illinois. April 15: Whitehall, Arkansas. April 16: Dorena, Missouri; Ware, Illinois; Wolf Lake, Illinois. McClure, Illinois. Then on April 21, came the big one, the kind of crevasse that everyone had feared.

Mounds Landing was a small village just north of Greenville, Mississippi, pressed against a sharp bend in the river. More than 22 million gallons of water hurtled past the town every second. Early that morning, a Thursday, even with sandbags adding a foot and a half of extra freeboard to the levee, the ice-cold water began spilling over the top. "We can't hold it much longer," one of the levee workers shouted. Then another shout: "There she goes."

This was not a small crevasse. The water formed a battering ram a hundred feet tall and thousands of feet wide—more than twice the water that typically runs over Niagara Falls, more water than ever flows through the

Upper Mississippi in the worst of its floods. When the Mounds Landing levee burst, tens of thousands of cubic yards of soil boiled into nothing, just more mud within the maelstrom.

The blast of water tore loose trees and barns and cabins. It carved a furrow into the earth a hundred feet deep and a half-mile across, pointing like an arrow away from the breach. Workers fled onto barges. The towboat strained against the force of the water and managed to escape to the Arkansas shore. That was a small bit of good news. The bad news was worse, though: a few hours after the breach, an engineer sent a telegram to Edgar Jadwin, the army's chief engineer, declaring that the flood would soon "overflow [the] entire Mississippi Delta." Indeed, the water eventually reached a hundred miles south and fifty miles east. It stood in places twenty feet deep. For decades, the drainage districts had talked about "reclaiming" land; now the water reclaimed. There was no longer a walled territory protected by a levee, just a river and its mud, expanding across its domain.

The sound of the water flowing over the landscape was like the wind, people said, like a tornado, like the snarl of a wild beast. As it reached into Greenville, the flood was, as one resident put it, "a glittering, slimy mass wriggling along the street like some horrid, merciless serpent." Dead cows and dogs and mules soon drifted atop the water, alongside tree branches, railroad trestles, and the remnants of shattered homes. Mud was everywhere. And bodies: the bodies of perhaps hundreds of Black men, farm laborers who had been marched to the levee, shackled, and forced to stack sandbags. They had been swept away in the explosion of water. How many died? Nobody bothered to count.

And still the rain continued.

After the Mounds Landing breach, the water slammed downstream, hammering through the levee in another half-dozen places over the next two weeks. A fleet of seaplanes fanned above the floodwaters, scouting boat routes, noting the creeping progress of the water. The pilots tossed out notes that were stuffed inside women's stockings and weighted with mechanical bolts, warning farmers that the water was bound for their land. Residents scrambled to help their neighbors. The islands between Mississippi and Arkansas were known to be wild and lawless, a shantyboat domain—"the corn-liquor center" of the nation, in the words of one plantation matriarch, a hotspot for bootleggers making illegal whiskey at stills hidden in the woods. Now the bootleggers put their boats to work as a relief army, zooming across the flood-soaked alluvial valley, rescuing stranded farmers.

Newspapers tended to headline the mayhem: men skulking across the river with explosives, aiming to protect their own land by blasting holes in the levee to flood the far side. That meant guards patrolled the riverbank, too. In early April, near a timber camp along a tributary in Arkansas, four men were shot for planting a hundred sticks of dynamite. In Greenville, the presence of more dynamiters led to what the local newspaper called a "pitched battle." Downstream of New Orleans, another man was killed because he came too close to the levee in his boat.

In late April 1927, after seeing the devastation at Mounds Landing, the Mississippi River Commission finally changed its tune: its leaders advised Louisiana's governor to save New Orleans by dynamiting a levee downstream.* This was not a carefully planned act of engineering—it was more like government-sanctioned sabotage. For ten days, explosions rocked the levee, until the breach was massive. Nearly two million gallons rushed out each second. This act spared New Orleans, but the flood washed over rural St. Bernard Parish instead. The region's residents had headed north to New Orleans a few days earlier, in what one reporter called "a highway of humanity": ten thousand people in trucks and wagons, atop mules and oxen, even afoot, urging cows up the road. They were promised restitution for the flooding, but little ever came.

* New Orleanians were already wary of the levees-only policy: if they looked north, after all, they saw a long and walled-in river that had nowhere else to evacuate its load. In 1925 the commission begrudgingly granted the New Orleans levee district permission to knock down eleven miles of the wall deep in the delta, near the river's mouth. Work on the project, known as the Bohemia Spillway, was completed just months before the river rose. In the face of the great flood, officials decided that even more room was required.

By the end of that spring, the levees along the Mississippi and its tributaries had failed in more than two hundred places. In June, a second flood crest, caused by snowmelt in the north, poured through the still-open gaps. Water covered 16.5 million acres across seven states.

In terms of area inundated, the 1927 flood was no worse than earlier high-water years. A flood in 1882, in fact, covered far more land. The problem now was that so much faith had been placed in the levees, and so much had been built behind the wall. (Andrew Humphreys's first rival, Charles Ellet, had foreseen this future, bemoaning the "false security" of the growing levee system.) Nearly six hundred thousand residents were swamped in Mississippi, Arkansas, and Louisiana. Some stayed on the upper story of flooded homes, but more than half wound up in boxcars and tent camps perched on the levees or on hills at the edge of the floodplain. Trash drifted along the camps' muddy edges. Mules and rabid dogs wandered the muck. Refugees took their meals squatting and sometimes slept on the wet ground. Estimates vary, but if you include those who starved in these camps, the flood may have claimed a thousand lives.

The situation was particularly bad for Black residents, many of whom had been blocked from evacuating, forced instead into camps where they worked to fortify the levees. Those who failed to comply with orders received no rations. Armed guards patrolled the refugee camps, letting no one leave without permission. To receive a pass, you needed a planter to sign, affirming you were returning to work in his cotton field. In Washington County, as little as fifteen cents was spent on daily rations for each refugee, and the best food was reserved for whites. As one man put it, giving decent food to Black families might "teach them a lot of expensive habits and there was no sense in giving them anything which they had not had before."

Herbert Hoover, the secretary of commerce, officially oversaw this relief effort. He made tepid investigations into the abuses; the closest he came to admitting what he'd discovered may have been when he told a

New York Times reporter that some white composer should have written "a plaintive dirge" for the refugees to sing in camp. ("Many good negro songs were written by white men," he explained.) The press was fawning: "Hoover New Hero of the Flooded South," the *Times* announced. The reporter claimed to have spoken to a Black man named Uncle Eph, who indicated that "de n[—] thinks well of Mr. Hoover. Sho' would make a noble president." Still, what mattered most was beyond Hoover's hands: how the next flood might be stopped.

The fate of the river—and the fate of its residents—now, once more, depended on decisions made by power-hungry men in offices a thousand miles away. This time the role of the levees was little debated. Everyone could see that the river needed some of its old floodplain returned. And so in September 1927, when the Mississippi River Commission released its initial proposal for how to stop the next flood, it called not only for new and larger levees but for four "outlets" in Louisiana and Arkansas.

Frankly, I have trouble distinguishing one chief of engineers from another. Their portraits hang in rows in the army offices: several centuries' worth of white men in drab military suits.* I am sure that some were kinder than Andrew Humphreys, or more committed to honest science. Edgar Jadwin, however, was close to Humphreys's second coming. A veteran of the Great War against Germany, he was deeply resentful that the Mississippi River Commission, even in its limited form, persisted after half a century. Indeed, in the years before the flood, as the army ignored European advances in hydraulic modeling, the civilian engineers had once more begun to critique the army's science. Jadwin did not want to take marching orders from civilians. He was a major general, and he wanted to order the march. So when the commission offered its initial flood-control plan, with a budget of $872 million, he demanded an immediate revision, one with a lower price tag.

* Two Black men have served as chief engineer, both in the past quarter century. No woman has held the role.

While the commission's leaders went back to work, Jadwin assigned 150 army engineers to draft a plan of their own. His team baldly stole the commission's ideas, but they cut corners to trim costs, which Jadwin knew would appeal to his friend and ally, President Calvin Coolidge. He proposed a dramatic cost-saving device: local levee boards, rather than solely the national government, would help pay for the infrastructure. A final, brash flourish, made Jadwin's true aim clear: the Mississippi River Commission's status would be diminished to that of an advisory council, whose directives the Corps of Engineers could choose to ignore. Coolidge had never been a fan of the commission, which he saw as, in essence, a lobbying vehicle for powerful Southerners. That Jadwin had managed to lower expenditures was an added bonus. The president gave the proposal his quick assent.

The plan was debated in Congress that winter. When he testified, Jadwin was aloof and combative. "You do not expect us to accept any plan simply because you present it, and to shut our minds to any other thoughts?" one congressman asked. "Yes," Jadwin replied, "I think you ought to do it." At times, he was simply wrong. The data in his plan, for example, suggested that the Yazoo Basin had not flooded in the era before levees. "That is news to us," one of the region's congressmen replied. Nonetheless, in May 1928, after months of horse-trading and veto threats, Congress passed a bill. The news was greeted in New Orleans streets with brass bands. Even with its reduced budget, the Flood Control Act of 1928 authorized $325 million in expenditures—the largest single public works project yet pursued by the federal government, larger even than the Panama Canal. One of its congressional sponsors hailed the bill as launching "the greatest engineering feat the world has ever known."

Jadwin had accepted two major compromises. The first hardly bothered him: Congress, asserting (once more) that this would not set a precedent for other rivers, agreed to pay the entire cost of the project; all local contributions were waived, in recognition of the heavy costs already incurred. The second compromise was more stinging. A three-member advisory board would examine and clarify the differences between Jadwin's plan and the Mississippi River Commission's.

A month later, in a recess appointment that did not require Senate approval, Coolidge chose a new commission president, one who he knew would side with Jadwin. Within sixty days, the board had finished its work, making no adjustments to Jadwin's proposal. Soon afterward the commission began to close its independent offices. Employees were immediately rehired by the Corps of Engineers. A half-century after Humphreys's battle against James Buchanan Eads, the army was fully in charge—a triumph for Jadwin.

For the river, though, this shift hardly mattered. It was just the end of a petty human squabble—a matter of what nameplate appeared on the desk where the buck stopped. All the engineers of the era still agreed on the most important point: that the river was an enemy that had to be subdued.

Herbert Hoover was sworn in as president in 1929, having used his response to the flood as a selling point. A former mining engineer, he leaned into this persona as president: he scheduled his appointments at eight-minute intervals and declared that he was going "back to the mines," after taking his lunch break. During his campaign, he had called for more infrastructure improvements, though the first big success of his presidency had been years in the making already: on August 27, 1929, Lock and Dam No. 53 was completed on the Ohio River—marking, after nearly two decades, the completion of the slackwater project. The Ohio River now featured fifty-one dams, which created a series of "pools," whose navigable depth would not waver with seasons. Hoover joined the flotilla of boats cruising the river on a celebratory voyage and, in a speech in Louisville, struck a hopeful note. Before long, he said, these dams would be regarded like Henry Shreve's snag-pulling effort: an impressive breakthrough but not nearly enough. The demands on the river would keep growing, he figured, and new improvements would have to follow.

The next year Hoover signed a bill authorizing a similar project for the

Upper Mississippi. Despite the fact that just six years earlier, 260 miles of this river had been protected as a national wildlife refuge, now dams went in. The bottomlands filled with water. Thousands of men were at work along the Missouri, too, laying more mattresses in an effort to finally tame the channel.

Down on the Lower Mississippi, the work now focused on implementing Edgar Jadwin's plan. Thirty miles upriver of New Orleans, crews were building weirs and levees for the Bonnet Carré Spillway. This is a tract of wetlands lined with embankments that run away from the Mississippi. Along the spillway's riverfront edge, the levee was removed, replaced by a "control structure": a mile-and-a-half-long concrete wall that consists of 350 separate bays, each one filled with vertical timbers. When the river grows too powerful—passing more than 1.25 million cubic feet each second—the timbers can be plucked free, one by one, to send some of the water through the spillway, into Lake Pontchartrain and safely away from New Orleans. Far to the north, the Birds Point–New Madrid Floodway was meant to protect Cairo. The floodway consists of 133,000 acres adjacent to the river; the levee at its upstream end can be dynamited open during floods so some of the river water can flow in, providing extra storage space. At the southern edge of the floodway, another opening can be blasted, allowing the water to drain back into the river's main channel.

Local landowners objected to these projects, but the Corps of Engineers won in court, so construction on both began by October 1929. The challenge was steeper in Arkansas and northern Louisiana, where Jadwin hoped to build the most important component of his system: two overlapping outlets that together would be nearly ten times larger than the Missouri floodway and over 150 times larger than the Bonnet Carré Spillway. More than fifty thousand people lived within the designated land, and the discovery of oil nearby had made their properties quite valuable. In 1929 one landowner filed a case in Louisiana court. A judge granted an injunction, halting work until the case was heard. Hoover caved to the pressure, and delayed construction. Hardly a year had passed, and already Jadwin's plan was out the window. Another ten years elapsed before Congress could settle on an approach that the politicians all deemed acceptable.

~

A stretch of river above Greenville, Mississippi, was once known as "Lazy Man's Reach." The channel here looped back and forth for twenty miles, in a series of horseshoes that turned the land into a stack of peninsulas. Together these bends provided a method whereby a paddler could use the downstream flow to actually *ascend* the river: if you put your boat in the current and let it drift a few miles down a bend, then dragged the vessel across a narrow strip, you'd find yourself upstream of where you began. Albert Tousley, a newspaper reporter who traveled the river in 1925, reported that an old fisherman used this method to return to his cabin after drunken escapades in Greenville: "Four 'drifts' downstream, about five miles each, four relightings of his pipe, four portages totaling less than two miles, and he had descended the Mississippi twenty miles to find himself forty miles above Greenville."

For the engineers, these bends were a concern—and not just because they slowed upstream steamboats. If the river ever punched through one of the narrow necks of land, the bed in the new shortcut would become a submerged waterfall, a sharp drop that might threaten the integrity of the nearby levees. For decades after Shreve's ill-fated 1831 experiment, the Corps of Engineers had a clear mission at every horseshoe: hold the river in place. This arduous, expensive work required the agency to continuously lay woven mats of willow boughs onto the banks to slow erosion.

Then in 1930 an army engineer named Harley Ferguson sent a brief memorandum to his superiors saying all this work was foolish. The better approach was not to fight the river's tendency to make cutoffs. The Corps of Engineers should go ahead and make the cuts itself.

Ferguson was a North Carolina boy grown into engineering genius. While stationed with the Corps's South Atlantic Division, he obsessively studied maps and charts of the nation's big river—the river that had turned the Corps into a hulking giant. And he liked to talk about the Mississippi as a machine. Its channel was a "pipeline," he said, its current a "dredge." His key idea was a riff on the jetty concept that James Buchanan Eads had endorsed half a century earlier: seize control of the machine by straightening the pipeline, hastening its erosive powers.

The army's engineers had a new tool at their disposal: a laboratory called the Waterways Experiment Station. This was another product of Jadwin's drive for power. One of the knocks against the Corps in this era was its lack of data, since the records on the river went back only a hundred years. Civilian engineers suggested that building hydraulic models, a new method being developed in Europe, could help fill the gap. Jadwin did not care much for models, but he knew that if Corps did not build a laboratory, then someone else might—once more threatening the agency's oversight. So in his 1928 plan, Jadwin granted the Corps of Engineers the needed authority. Now his successor, Lytle Brown, set the engineers to work building an eighty-foot-long scale model of the Greenville Bends inside the new laboratory.

Assembled inside a warehouse in Vicksburg, this model was an intricate machine. To gauge the direction of the current and its eddies, the engineers tied threads to the banks and poured colored dyes into the water. To determine the behavior of river mud, they used creosote-soaked sawdust as a proxy, dumping it into the model channel and seeing how it moved. Months of experiments offered mixed but largely promising results for Ferguson's river-straightening scheme. So the engineers moved on to a more extensive study on a larger, outdoor model.

Brown, like Jadwin, didn't care much for this tinkering. He was ready to press forward. "The best that the laboratory can do is to give indications," he told Congress, asking for permission to get out and dredge the river. Brown got his permission, promoted Ferguson to brigadier general, and put him in charge of the Mississippi River Commission.

Their 1932 meeting in Brown's office, retold in Corps of Engineers legend, is perhaps too perfect to be true. The two sit chatting, smoking their pipes, until Ferguson grows impatient. "Do you want me to write a book or fix a river?" he asks.

"Fergie," Brown replies, "you get the hell out of here and go fix the river."

As one of his first acts, Ferguson ordered the river straightened at a bend called Diamond Point, just downstream from the Waterways Experiment Station. The plan was to hasten a natural cutoff that had begun developing a few years earlier. (Once the concept was tested at Diamond Point, Ferguson would work northward, carving more cutoffs, gradually preparing the river for the momentous change that would be the full elimination of the Greenville Bends.) The Corps set two dredgeboats to work slicing a two-and-a-half-mile line through the thin neck of this fifteen-mile bend, one headed north, one south. After a few weeks, a thin wall of earth remained between the two new channels. On January 8, 1933, the water on the upstream end stood five feet higher than the water downstream. At ten a.m., amid a light rain, the engineers dynamited the plug, sending dirt and clay flying upward. Workers ran in to smash away the last of the soil. Within hours, what had started as a small hole had widened into a gushing channel, sixty feet across.

There was little fanfare in the press. "Nobody even noticed much the shot that should have echoed up and down the Mississippi Valley," a New Orleans newspaper wrote later. Nonetheless, a new era had begun. The Corps of Engineers had launched the largest river channelization program ever undertaken on the planet.

Perhaps no one noticed because they were distracted by suffering: four years earlier, as the battles over Jadwin's plan were still raging, the Great Depression had arrived.

Now even more shantyboats appeared on the river. One report in *The Saturday Evening Post* estimated that fifty thousand people lived in boats moored along the Ohio and Mississippi. In New Orleans, permanent homes were built from salvaged barge wood; houseboats that had been floated down to the city were reinstalled on land. These "batture dwellers" caught shrimp and catfish, which they ate themselves or sold for profit. They kept pigs and ducks and chickens. Such people did not care about a straightened river. They wanted money. They wanted jobs. So did millions of others across the country. And Franklin Delano Roosevelt promised to deliver jobs and money, which was how he won the 1932 presidential election in a landslide.

It was Roosevelt's new generosity that truly remade the Mississippi watershed. Hoover had signed off on a plan for twenty-four locks and dams on the Upper Mississippi, but Congress never sent any money. Now a tranche of cash went north. Thousands of laborers settled into work camps, setting temporary dams into the channel so they could drain the water where they needed to work. They drove steel piles and dropped concrete trestles into place by crane. These new locks would be good for river navigation, sure, but Roosevelt's aim was as much to maximize the number of government-supplied jobs.

Roosevelt signed a new Flood Control Act in 1936, which specified that stopping inundation was a "proper activity of the federal government"—not just on the Mississippi but on *all* rivers. This was one more boon for the Corps's engineers, who had a rush of new projects to build. Hundreds of dams went up over the next few decades. None were built on the main channel of the Lower Mississippi, which was too mighty a torrent to be contained within a reservoir. But on its obscure back tributaries—the Fourche La Fave River in Arkansas; the Tygart River in West Virginia; small creeks from Illinois to Oklahoma—dams were built and reservoirs began to pool.

Ferguson, meanwhile, pushed forward with his dredging. He was not yet sure whether the Diamond Point cutoff had done anything to reduce flooding, but shortening the channel had caused no catastrophes, and that was proof enough. In the spring of 1933, the Corps of Engineers had carved two more cutoffs just downstream of Diamond Point. By 1935, the

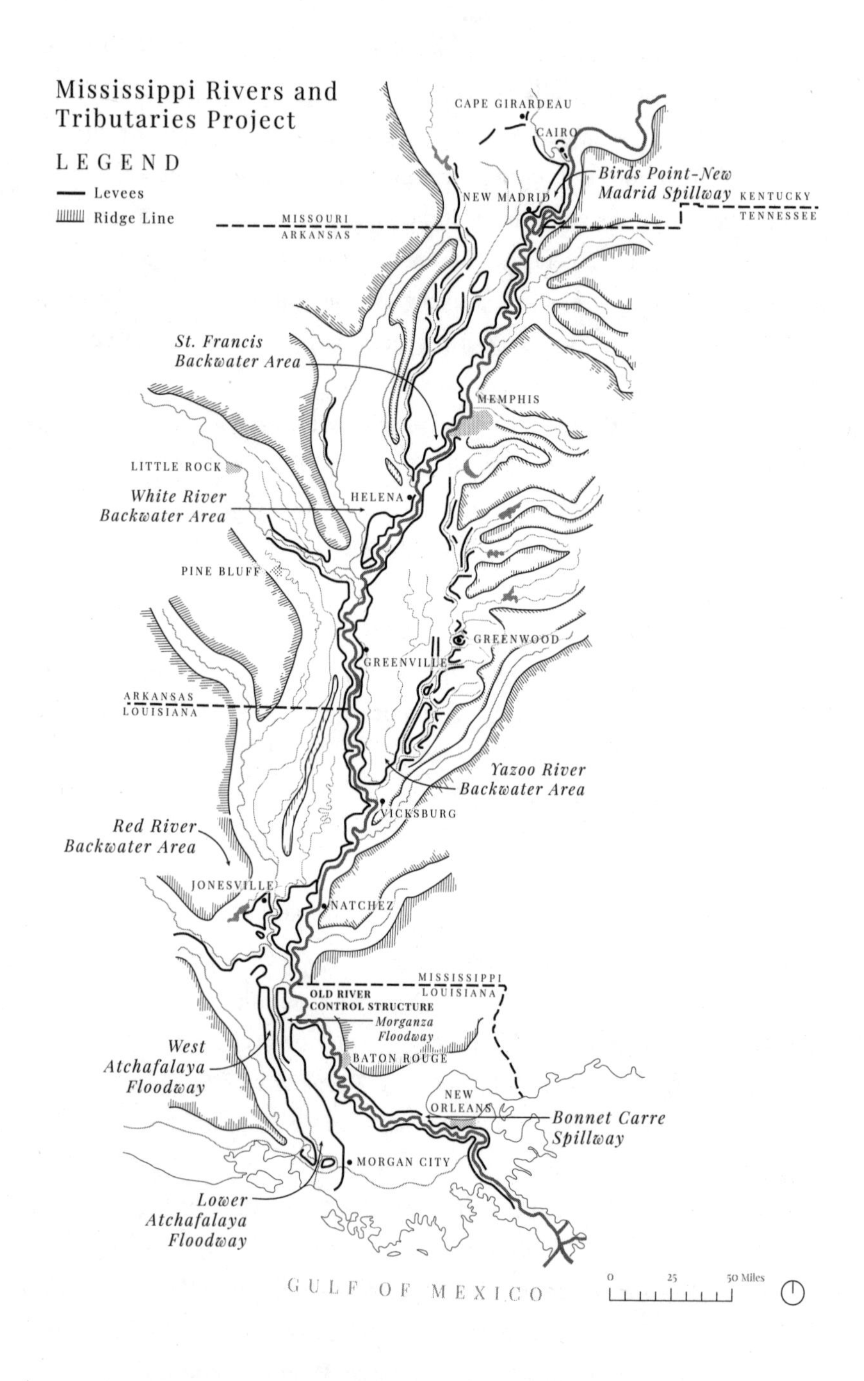

Major components of the Mississippi River and Tributaries Project.

Corps had tackled the mighty Greenville Bends. (They did it sooner than planned, in fact, because the river had carved its own cutoff across one bend.) The Mississippi had shrunk by ninety miles. A 1937 flood suggested the program's promise: the river now carried more water at a lower stage.

This success opened new possibilities. Because flood heights were dropping, the contentious safety-valve floodways in Louisiana and Arkansas would not need to hold as much water. So a series of laws revised their layout, which was finalized in 1941. The new plan skipped over Arkansas entirely—the state would feature no floodways—and focused instead on the Atchafalaya River. The West Atchafalaya Floodway would allow water from the Red River to bypass the Mississippi River during floods; to the southeast, the Morganza Floodway would divert Mississippi water away from Baton Rouge. The engineers installed a row of 128 concrete structures, each framing a steel gate, at the spillway's inlet. The water from both floodways would pass down the Atchafalaya, which offered an alternate route to the Gulf of Mexico.

By the time the work was completed at Morganza in 1954, more than two hundred dams dotted the Mississippi's tributaries. The great river itself had been shortened by nearly 150 miles. Engineers were building more wing dams to constrain the river; they were protecting its banks not with willow branches but with layers of concrete. In fits and starts, in other words, with no real central plan, a new watershed was being built.

The Corps of Engineers eventually settled on a name for the sprawl of infrastructure along the southern river: the Mississippi River and Tributaries, or MR&T, Project. Despite its hodgepodge creation, its extent makes it "among the largest and most ambitious engineering feats on earth," in the words of Paul Hudson, a hydrologist who has written a textbook on river engineering.

I've visited all the MR&T Project's flood outlets. I've gazed upon the massive concrete fortifications that guard their thresholds, and I've driven through the low-lying land that has been walled off to contain the occasional flood. Waving with soybeans, the land inside the floodways looks

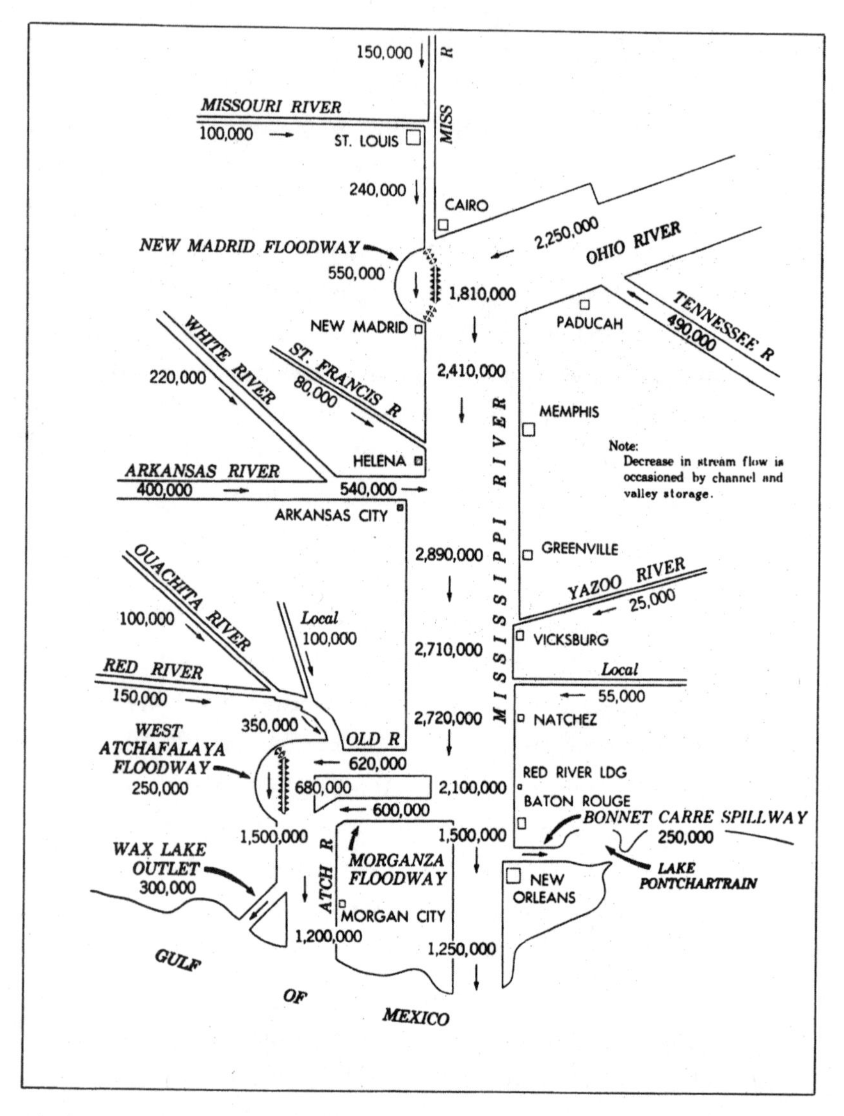

Discharge rates, in cubic feet per second, of various components of the MR&T System.

little different from any other stand of former floodplain, just slightly emptier of anything human-built. I have watched at the Bonnet Carré Spillway as the timbers were pulled upward, releasing the continent's outflow. (The water crashes down into the spillway in a miniature waterfall; birds like to hover, snapping up the fish caught in the flow.) I have surveyed the levee from Louisiana to Illinois. Sometimes the hump of earth is topped by a paved path, sometimes gravel, sometimes flattened grass. I find it hard, on these various tours, to conceive of each of these pieces as linked to all the others.

One way to grasp the extent of the project is to consult the seemingly endless cartographical portfolio assembled by the Mississippi River Commission. Maps of the outlets and engineering diagrams depict the dimensions and configurations of their gates and towers, and charts of the backwater areas in Arkansas and Mississippi indicates where floodwaters are allowed to push upstream at the mouths of smaller tributaries. One map's thick red lines trace the 3,800 miles of levees—2,200 on the Mississippi and its sole remaining distributary, the Atchafalaya, plus another 1,600 on backwater tributaries—that were expanded and improved as a part of this project. Another renders the river not as a winding waterway but as a branching set of pipelines. Each pipe corresponds to an incoming river or an outgoing floodway, and its width signals how much water the levees and gates are designed to accommodate. These numbers were not chosen at random. The Corps of Engineers has carefully calculated a hypothetical apocalypse. The latest version, updated in 1954, envisions what might happen if the rainfall from three historic storms—in 1937, 1938, and 1950—arrived over a matter of days. The resulting flow is known as the "project design flood," and the spillways and reservoirs and levees are all arranged to accommodate its rush. The map is a triumph of geometric order, suggesting that the river has been fully put under control.

But even the Corps of Engineers has had trouble understanding the whole of its work on this great watershed. New policies in this era required the engineers to conduct cost-benefit analyses to justify each project, but these studies were local. How did the individual projects affect places up- and downstream? Nobody knew. The reservoirs were particularly vexing: how do you time the release of water from so many dams so as to avoid any flooding when they merge downstream? That's why, in 1943, soon after the MR&T Project's layout was finalized, the agency decided to build the largest hydraulic model in the world.

The Mississippi Basin Model, as it's officially known, includes nearly all of the Mississippi Valley. There's the Missouri as far as Omaha; the Arkansas to Tulsa; and the Ohio River Basin out to Louisville, including portions of the Wabash, the Tennessee, and the Cumberland. Strangely, the Mississippi River itself reaches an artificial end at Baton Rouge, hundreds of miles short of the actual mouth. The Corps claimed that, since no more tributaries entered beyond this point, the river in Louisiana didn't need to be studied. Nonetheless, they did include the whole of the Atchafalaya River, since that distributary was undergoing its renovation from natural swamp into carefully controlled floodway. Even shrunk by a factor of 2,000, this model covers more than two hundred acres, an area larger than Disneyland.

In order to ensure the river was not a shallow sliver, the vertical scale had to be skewed and exaggerated; this meant that from the top of the model watershed to the mouth of the Atchafalaya, the altitude within the model drops fifty feet. The Waterways Experiment Station did not feature the needed terrain, to say nothing of the space required. So the Corps of Engineers acquired eight hundred acres in the rolling, wooded hills in central Mississippi, forty miles east. Since the nation was at war again, the army's engineers had a cheap supply of labor: German and Italian prisoners. By the end of 1943, the work site, known as Camp Clinton, was home to nearly two thousand captive soldiers. The camp was secured by two parallel lines of woven-wire fencing, ten feet tall, topped with loops of barbed wire. Armed guards in towers watched for escapees. But not everything felt carceral. The kitchen had access to higher grades of beef and butter that, due to rationing, were off-limits to U.S. citizens; the food was so good that visiting staff would time their arrival to coincide with meals. By the time the war was over and the prisoners sent home, they had moved a million cubic yards of dirt, clearing six hundred acres of forest.

Now the model was installed. Every bridge and levee, every highway ramp and railroad fill built along the tributaries was rendered in miniature. Trees were recreated with wire screens, folded to match the average height of local forests, placed based on data collected through aerial photography. To mimic the rough surface of the river's bottom, the Waterways Experiment Station conducted tests on a smaller model, searching for materials that would provide suitable drag. These studies suggested that the model riverbed needed to be carefully scored in places; angular brass plugs should be installed elsewhere, their lengths varying to create the needed friction. The real-world landscape was sectioned into a grid, with each square to be recreated on a block of concrete. These blocks—each measuring 150 square feet—were laid in place at precise angles, calculated to reflect the tilt of the earth. As the model came together, other engineers developed new automated sensors, which helped ensure that rather than a staff of six hundred workers, the model could be run by sixty men.

This concrete wonderland is, in a sense, the logical conclusion of the process that had begun more than a century earlier, back when Prospect Robbins and Joseph Brown set their "initial point" deep in the Arkansas swamps. From that axis, generations of surveyors had established their lines; the land was charted, then sold, then tamed and cleared. Now in the model, it would be reproduced completely and entirely, its shape perfectly matched, so that it could be finally and fully understood. Etched into concrete, the river was frozen in time. Perhaps Jorge Luis Borges had this project in mind when he wrote his fable about the map and the territory, which was published in 1946, just as work on the model kicked into high gear. Certainly the story seems prescient now: this map, too, has been reduced to tatters and waste.

Construction was supposed to last a year, but nature posed problems. The workers discovered that the local clay was so soft that the concrete panels heaved unpredictably. Rainfall leaked through the joints in the grid. The model had to be realigned every month. The world is messy, it turns out, far messier than a plane of concrete.

An even bigger problem was the intermittent funding from Congress.

With its budget limited, the Corps decided to build in pieces, so they could use parts of the model even before the whole system was complete. This proved a wise choice; the model's great triumph came in the spring of 1952, as a major flood rushed down the Missouri River, whose miniature twin had already been built. For sixteen days, the team in Mississippi worked around the clock: programmers in the control houses initiated on the model flood; the automated recorders spat out the data. Daily phone calls sent the gathered intel—water-surface profiles and discharge hydrographs—to army engineers in Omaha and Kansas City. By clarifying which levees could be raised to stop the flood and which were a lost cause, the model prevented an estimated $65 million in damage.

But by 1959, sixteen years after the plan was first conceived, the concrete river reached only as far as Memphis. And by the time the work was finished, in 1966, the thing was all but obsolete. Computers were emerging, and so this model, two decades in the making, was now used to test and tweak the digital models. Then, three years after construction concluded, the Mississippi Basin Model was retired. The Corps of Engineers decided to leave the thing open to tourists—a riverine theme park. Finally in the 1990s, the federal government tired of paying for upkeep. They deeded the property to the city of Jackson, which converted the surrounding land into what is now Buddy Butts Park. The model was locked behind a chain-link fence.

Today, on the fence outside Butts Park, four separate signs in different phrasings repeat the same basic message: GATES LOCKED AT DARK. Styrofoam cups and empty PBR cans and black trash bags litter the grass. Boastful, self-granted nicknames are spray-painted across the concrete, along with the occasional oversize phallus. This park lies forty miles from the Mississippi River, on the very edge of the watershed; just two miles east, rainfall follows a different river to the ocean. But to me the concrete model feels like the soul of the modern Mississippi, the river of these American years. There are gaping holes in the fence around the model, and if you push them aside, you can walk unbothered into an overgrown world. Weeds—sometimes whole trees—grow up from gaps in the concrete. Back corners, never paved at all, are covered in young forest. The concrete river channel disappears under drifts of rotting pine needles.

Years into my study of the Mississippi, after I had paddled the entirety of the lower river, I visited the model, hoping to conduct a survey, walking through the various sites I'd seen. Instead, confused by the chaos of vegetation, I found myself lost. I had no idea whether the trough beneath my feet was supposed to be the Ohio or the Missouri or the Upper Mississippi. The forest was so thick that I'd lost all sense of cardinal directions. Which way back to my car? I was unsure.

Eventually I found a watchtower, a relic from the model's years as a theme park. I trudged to the top, climbing steps that were streaked with rust and littered with cans and bottles. My salvation, I thought: from above, everything would become clear. But during the years since the model-turned-theme-park was abandoned, seeds had settled, then sprouted into saplings. Now the trees stood as tall as my watchtower. All I could see was a mass of leaves—an intrusion of life, paying no heed to the concrete map.

Part III

THE UNMADE MISSISSIPPI

TODAY (AND TOMORROW)

Exposed concrete revetment during low-water conditions (Washington County, Mississippi).

9

Hooked Up, Hard Down

The uncertain state of the big rivers

I've come to realize that I'll never know this river as well as the fish do. From their point of view, down in the blackness of the water, where mud chokes out all the light, the Mississippi watershed unfolds into a vast universe. The water flows atop sand and silt and rock, through channels big and swift and, in some seasons, up oxbow nooks that have been reconnected by the floods. These thousands of miles of water have provided space for refuge throughout many eras of change. When the glaciers pushed south, the fish could flee into the Lower Mississippi; when the ice rolled back north, the fish could follow. That means that today the watershed's many branches form what scientists have called "a primary center of diversity of the North American fish fauna"—home to 260 different species, almost half of the continent's total. You cannot know the river's story without knowing theirs. After all, some of these species' presence in the watershed precedes not just ours but the Mississippi River's itself.

Seventy million years ago, when the crumpled trough that became the Lower Mississippi River was still a bay, an even larger inland sea reached across the Texas prairies, north onto the Great Plains. Large fish lurked in the warm and shallow water that lapped against the foot of the Rocky Mountains. These fish—of the family Acipenseridae—were scaleless, their bodies clad in a series of interlocking plates. They hovered at the muddy bottom and used their mouths like vacuums to suck up nutrients. Some of the world's best-preserved Acipenseridae fossils are from this era and were found in the dirt alongside the tributaries of the Upper Missouri, in Montana. Their descendants, little changed, remain in the Upper Missouri today—one small slice of the 3,500 miles of the Mississippi and its tributaries where one modern species, the pallid sturgeon, now ranges.

These fish have seen so much chaos, so much change. The seas dropped and new rivers appeared, yet the sturgeon stuck around. The water snaked and meandered and built new valleys, and Montana's canyons became linked to the network of water that drains to the Gulf of Mexico, but the sturgeon survived. The Mississippi Embayment filled. The earthworks appeared. Empires fought and lost. Snags were removed. Steamboats rose and fell. Still the sturgeon remained.

Then, in the middle years of the twentieth century, a series of laws tasked the Corps of Engineers with creating shipping channels at least nine feet deep on the Mississippi River and its major tributaries. So dams were built, and the Lower Missouri was "channelized"—straightened into a river so fast and narrow it could self-dredge its channel. Suddenly, in less than a geological instant, within the half-century lifespan of a single pallid sturgeon, its universe changed. As many as 150 square miles of riverine habitat were disconnected from the eight hundred miles of the Lower Missouri. The fish began to disappear, too. According to one study, by the time you read this book, wild-born pallid sturgeon may be gone from their ancient home on the Upper Missouri.

Biologists have been searching for decades for ways to save this species. The best current theory suggests that the trouble starts with its larvae: they have no fins, and so they get caught in river currents, which at the new cranked-up speeds give the larvae no chance to settle in the slow backwaters where they best feed. The proposed solution is the installation of interception-rearing complexes (IRCs), small tweaks in the river dikes that create off-ramps, steering the larvae toward the river's slower-moving edges. In 2018 the U.S. Army Corps of Engineers—which has, through the years, expanded further and now engages in environmental restoration—committed to building twelve IRCs, gradually, over six years. But a coalition of agricultural and navigation interests worried that their installation would shift the flow of mud and build new shoals.

Robert Jacobson, the U.S. Geological Survey hydrologist who developed the IRC concept, is skeptical. "You could take most [people] down the river, they could pass an IRC and never know it was there," he told me one summer afternoon, as we roared up the Missouri in his research vessel. Two hundred miles above St. Louis, the river is lined by forested limestone bluffs and has its trademark murk, stained by the loose soils of the Great Plains. I was, per Jacobson's instruction, watching for a notch in a bit of rock in the middle of this dark water. But despite his insistent pointing, I couldn't distinguish anything.

In the hour we were out on the water, no other vessels appeared. Indeed, I spent two weeks driving up and down the Missouri River and never saw a towboat. Still, the concerns of the pilots won out: just two IRCs were built before a Missouri congressman added language to a massive spending bill, forbidding any further construction until the Corps of Engineers conducted a full study on their impacts—whether they'd impede navigation or cause floods on nearby farms. To Jacobson, this felt like a catch-22. The first twelve IRCs were supposed to *be* the experiment that could inform such a study. The Corps is legally obligated to protect the pallid sturgeon. Now it was blocked from employing one of the most promising strategies to save the sturgeon's life.

Perhaps because this river was for so long the path out into the Wild West, I could not help but think of the old cowboy saying: *This town ain't big enough for the both of us.*

A hundred years ago it was possible to accuse the navigation boosters of, as one editorial put it, psychopathic enthusiasm. But the enthusiasts won the argument. Within a decade of the completion of the Ohio River's slackwater system in 1929, traffic was outstripping the Corps of

Engineers' initial projections. By the mid-1950s, so much coal and steel and petroleum was being carted downriver that the system needed to be "modernized"—the old dams removed, new dams installed, the locks expanded. It's an ongoing project that has continued, piece by piece, ever since. The Upper Mississippi was fully fitted into its own system of dams by 1940, and over the next two decades, local traffic increased tenfold, to 27 million tons, then kept growing from there. The rivers were no longer the moribund backwater that Mark Twain had encountered in 1882 but, once more, a network of nationally important highways.

In the 1960s, the "towboatingest" town on this system was Greenville, Mississippi, a small city of 47,000 that featured twenty-six companies building or repairing or managing boats. The rest of the Yazoo Basin was already past its prime, beginning to depopulate. Greenville's sheltered harbor, though—built into one of the old riverbends that had been abandoned by Harley Ferguson—bustled with ships that were being built and filled and cleaned. L. D. Williams grew up in Greenville, though he eventually left for an office job at a mortgage firm downstate, in Jackson. When he grew sick of his boss, he remembered what he'd seen as a kid. So he quit the job and called a friend, a bargeman, to ask for work. After waiting two weeks, Williams called again. His friend was surprised to find that he'd been serious. "He said, 'Let me tell you something, this is hard work,'" Williams remembered later. "And I said, 'I can do it.'"

The friend provided a packing list—five shirts, five pairs of jeans, steel-toed boots, underwear, socks. Williams filled his bags and stepped on board. After that precipitous turn, his career ran relatively straight. From deckhand he became mate, then engineer, then first engineer. Williams reported to the captain—"the wheelman," he called him—but in most circumstances his own authority was absolute. Such is the path of any successful towboat career; even a captain must "come up through the deck," as the saying goes. They make good money, a wheelman especially. "A hundred and twenty, a hundred and thirty a year," Williams said.

We chatted in his second-floor apartment—once the site of a brothel, he told me, that, given its location just blocks from the levee, would have serviced his towboat predecessors. Williams spoke in a loud voice. I found

I had to use a loud voice, too. Thirty years of sleeping in a cabin above a grinding engine had taken its toll. Williams had recently retired, in part because new restrictions forbade him to smoke on board the boats. He was razor-clean bald and short and appeared to be built entirely out of muscle.

To me, one allure of a long trip down the Mississippi River is the clean sense of accomplishment: you launch from one port and head to another—downstream, ideally, if you're in a canoe—tracking the sights along the way. Williams described a more formless kind of voyaging. Once his boat's cargo was dropped in a port, the next shipment might lead the crew downriver or farther upstream; you never knew until new orders arrived. The boat that Williams worked his first year plied the same waters again and again, never reaching beyond Rosedale, just fifty river miles upstream.

On a towboat, then, a journey is best measured not in distance but in days. Once on board, Williams worked for thirty straight. Then he took fifteen off. "That's your life right there," he said. "You live out there more than you do at home." Which meant no pets and few landside relationships. Such was his life for close to thirty years.

Nonetheless, Williams described an appealing existence. The crews ate well: per towboat tradition, there was fish on Fridays, steak on Saturday, chicken on Sunday. "I did all my work first thing in the morning," Williams said. "I did it and I did it hard." Every fifteen minutes he had to complete a walkthrough, but otherwise his days were spent sitting in an air-conditioned office, reading—up to fifteen books each month, which he packed into a separate library of a suitcase. Some of his peers were not fit for this life. One man objected to what he called "woman's work": all the toilet-scrubbing and bed-making that takes up so much of a deckhand's day. That man came on board in Greenville; he was off by Lake Providence, Louisiana, the next port, fifty miles downstream.

These boats are a far cry from the old steamers. After the Civil War, the boatyards began cranking out a new sort of vessel: an updated flatboat with the living quarters removed. "Barges," as these became known, could be hitched to the steamboats, expanding their capacities. A few decades later, the iconic paddlewheels began to disappear, replaced by screw propellers. Then diesel engines succeeded the old steam boilers. With no need to carry passengers—who after the war, like everything else, began to travel by rail—the boats themselves shrank down to the bare essentials: a pilothouse, an engine, plus quarters and a kitchen for a crew. The modern towboat is a squat little beast, snub-nosed. Add the barges, though, and a single vessel becomes huge—longer, sometimes, than the rivers are wide.

There is a statistic often touted by this industry: a single barge can carry enough grain to bake 2.5 million loaves of bread. Since a towboat on the Lower Mississippi River can move more than fifty barges, one captain and his crew can cart 125 million future loaves, the beginnings of well over a billion sandwiches. All this cargo, I should note, looms before the pilot in a long skinny line, delving into water he has not yet met: a towboat, despite the name, does not tow. These boats push their cargo. Which, given the hydrodynamics, can be a scary fact. As one retired captain, David Greer, told me, a towboat *has* no hydrodynamics. It's a giant raft, built to hold as much cargo as possible with very little draft. "And you don't navigate a raft," Greer said. "You position it in the current." When a pilot is running too fast through a bend, he might have to go "hooked up, hard down," he added. "In other words, you've got your rudders as far over as they'll go and your engines full ahead. You're trying to get up off that bend and stay up off it." Greer sees this as an act of desperation.

Greer, like Williams, is a bookish guy. He took some time off piloting to study history at Louisiana State University, but he soon decided that being on the river offered a deeper and more visceral connection to the past. He was part of a fraternity of men—it's men, mostly, who work these boats—who are the spiritual descendants of the shantyboaters, the self-named river people. (Several towboat pilots are the literal descendants of shantyboaters: born in a floating shack atop these muddy waters, they found a way never to leave.) The towboat crews are among the last to know the watershed in all its expansiveness, the last to remember the names of the islands and bends. These are the kind of people who like to say their blood is made from muddy water. Still, the routine of a towboat enforces a certain distance. These vessels never stop for the night; groceries and supplies are delivered by service boat. So if the river is already its own sealed world, clasped between the levees, these boats make up another, smaller world, adrift on the endless tide of water.

Such are the inner workings of this forgotten corner of the U.S. economy. Altogether, U.S. inland rivers carried $134 billion in cargo in 2019, creating half a billion jobs along the way. Supporters of the shipping industry point out that the dams create recreational opportunities, too, spawning lakes that are easily navigable and fishable by powerboat. The dams supply municipal drinking water by creating reservoirs; private companies have installed hydropower plants, generating energy. Most of all, the boosters point out that no other country in the world features a network of navigable waterways running through rich and productive farmland. Inland navigation is often described as safer and cleaner than railroads and highways and airplanes, given the low rates of emissions per pound of cargo.

Critics dispute this last talking point. Towboat efficiency is undercut by the river's "circuity," or its winding nature: compared to trucks and railroads, goods on the river need to travel farther to reach the same place. And the critics have plenty of other talking points, too, since not every navigation project has proved worth its cost. The Corps installed thousands of dikes along the Lower Missouri, narrowing a river that once spanned 2,400 feet to a torrent just 750 feet wide. Afterward, the water flowed three times faster. But Missouri traffic hit a peak in 1977, at levels below the projections that the Corps of Engineers had used to justify the project. That same year the Congressional Budget Office found that inland waterways were the nation's most subsidized form of transportation. Government handouts made up more than 40 percent of barge companies' revenues.

That report was one salvo amid the latest fight on the rivers. The Corps had grand plans to replace Lock and Dam No. 26, which spans the river at

Alton, Illinois, just upstream of St. Louis. The concrete was disintegrating. Scour holes ran along the riverbottom. The dam had become a crumbling old wall, ready to blow.

The Corps wanted to build a supersized new lock. Cargo loads had kept on growing, and for a pilot in a modern townboat, the river's original locks seemed tiny. The crews had to stop and break apart their barge trains, going back and forth to push everything through. At Alton, the Corps could take the first step toward an even better river system—a channel twelve feet deep that could accommodate even more cargo. By then, though, such ambitions faced a new obstacle: the environmental movement that was born in the age of Mark Twain and John Muir had, after a hundred years, reached its maturity.

The scale of the crisis had become impossible to ignore. Rivers were so soaked in oil that the water caught fire; birds were dying, poisoned by pesticides before they could emerge from their shells. The Mississippi's tributaries were seen, not incorrectly, as cesspools. For years, cities had seen the flow of their rivers as a convenient method to carry off waste: slaughterhouse refuse, the gritty effluent of steel mills and ironworks, even human feces. By the mid-twentieth century, once industrial chemical wastes were added to this slew, the water had turned toxic. Five million dead fish floated to the Mississippi's surface in Louisiana in 1963. As such horrors began to crowd newspaper headlines, Congress passed a wave of environmental laws.

Some of these laws—the Wilderness Act, the Endangered Species Act—are iconic. The National Environmental Policy Act (NEPA), signed on New Year's Day in 1970, was a bit less glamorous, but it's essential to the story of the river. NEPA requires that federal agencies complete comprehensive studies, examining the wider impacts of their projects. They have to hold public meetings to share what they find. The law, then, was one more assault on the Corps of Engineers' hegemony—not from rival engineers now, but from historians and archaeologists and biologists, people with very different ideas about rivers. A wave of lawsuits halted many of the Corps of Engineers' so-called "improvements," including Lock and Dam No. 26 in Alton. In the wake of NEPA, amid this legal chaos,

Congress failed to pass any major law authorizing new work on rivers and harbors for fifteen years.

What broke the impasse was a compromise. In 1986 the omnibus Water Resources Development Act authorized a laundry list of new navigation projects—but in a new wrinkle, it tasked the Corps of Engineers with environmental projects, too. The military engineers were entering the ecosystem business.

The Mississippi River earned its own full attachment. The Upper Mississippi River Management Act declared the river to be both "a nationally significant ecosystem and a nationally significant commercial navigation system." Officially, this "system" was to be "administered and regulated in recognition of its several purposes." So for the first time on a big U.S. river, a program of environmental restoration began. It was meant in large part to compensate for the damage that was occurring in Alton, where construction on the new dam was finally underway.

I wanted to see the results of this program, so on a journey along the upper river, I drove south into the Driftless. This region, which has been known for its beauty since even before the steamboat days, is named because its tens of thousands of square miles somehow escaped the wrath of glaciers; here there is no "drift," as glacial sediment was once known. Unlike the rest of the Midwest, the ancient hills remain, unleveled by the ice. They rise so steeply above the river that early explorers described them as mountains. Creeks have carved deep canyons that drain the forests; sheltered between the peaks, the river is lined by a low nook of marsh and wetland trees. Long before the 1986 compromise, then, this was a place of ecological concern: it was the beauty of the landscape that Congress hoped to preserve with the Upper Mississippi River National Wildlife and Fish Refuge. The refuge's creation blocked plans for local levee districts, but Congress made clear that its presence could not stand in the way of the navigators' needs. And just six years later, the federal government decided the towboats needed dams, which were the only way to make this shallow water reach the legally required depth of nine feet.

The official line was that this new work would benefit the refuge, too, and when the project was finished in 1940, locals found plenty to like.

Waterfowl hunting had been rare here; there was not enough slow and open water to supply the kind of food that attracts ducks. With the river now converted into a series of lakes, the still bottoms blossomed into vast red fields of smartweed. The seeds attracted waterfowl in record numbers. Muskrats boomed, too. Fish began to congregate just below the dams, feeding on whatever slipped through the gates. By some reports, commercial fishermen's income in the late 1950s was three times what they'd made before the dams were installed.

When I visited the refuge, it felt I'd arrived at a river entirely separate from the great southern waters I know. In Arkansas and Mississippi, it's rare to see any vessel but towboats; here, meanwhile, every village featured a marina lined with speedboats and yachts. It didn't take long to see why so many people cherish this scenic northern stream. Throughout the rainy night I spent in the Trempealeau Hotel, in Wisconsin, a train roared past every few hours, running along the edge of the water. The river formed a wide, still lake that reached to the foot of the surrounding hills—a lake that in my opinion deserved a richer name than Pool No. 6. To the south I could see the pool's precipitating source, Dam No. 6. This was nothing like the terrifying grandeur of the Hoover Dam, which holds back a wall of water seven hundred feet tall. The dam in Trempealeau was a humbler construction, just a line of low gates, which somehow reminded me of a brace on a broken leg, or a bit of scaffolding—some minor buttressing meant to keep this lovely lake intact.

But this placidity belies the hidden damage. Paddlefish and blue sucker, blue catfish and lake sturgeon—another threatened member of the Acipenseridae family—have all but disappeared north of Keokuk, Iowa, where the river is slashed by its second-tallest dam.* The skipjack herring is gone from this reach, too, and with it went the ebonyshell mussel, which depends on the herring to host its larvae. The worst effects are to the south of the refuge, where, in addition to the dams, levees have sliced away roughly half of the floodplain, cutting

* This dam actually precedes the nine-foot channel; it was completed in 1913, to fully drown the Des Moines rapids.

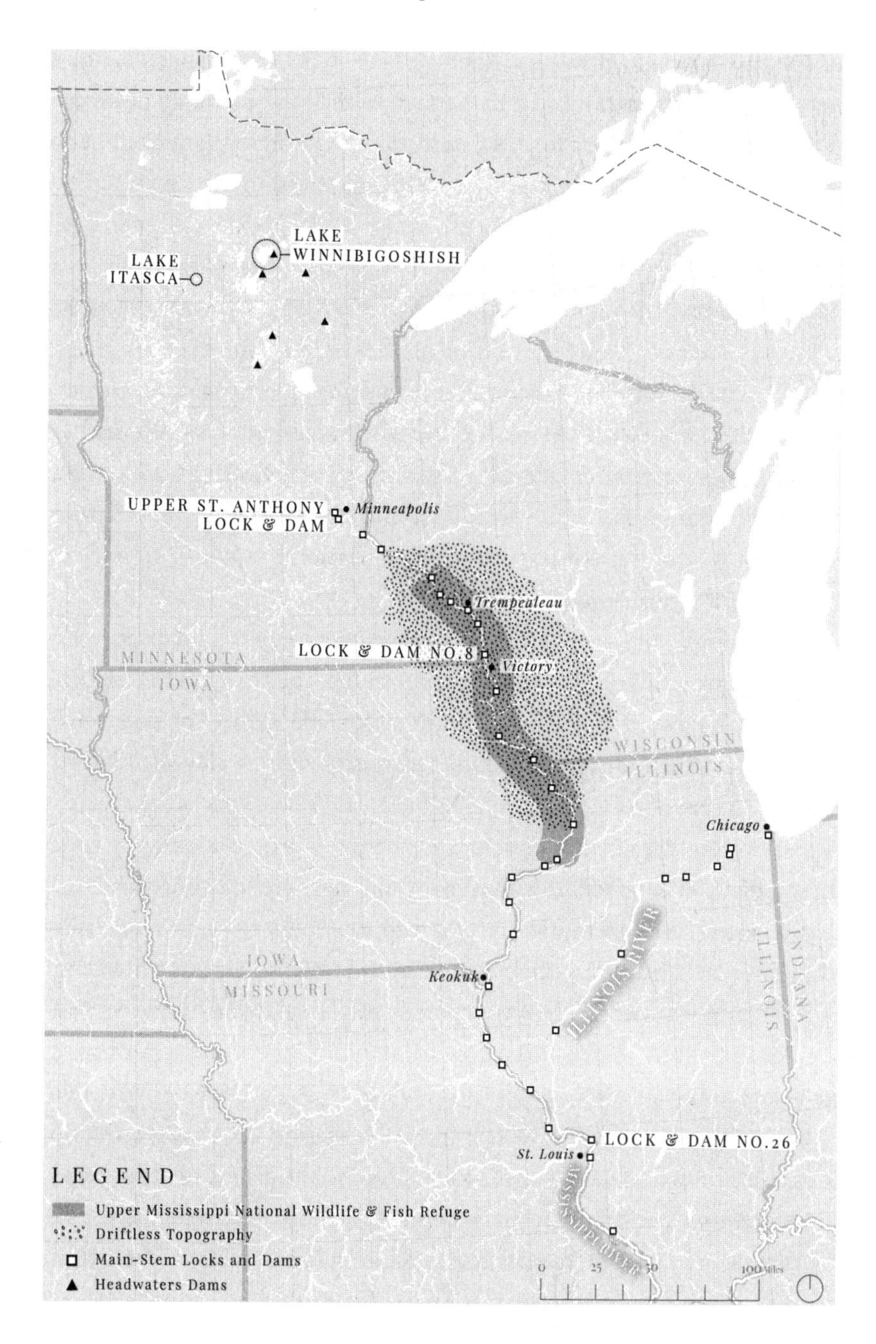

Key sites along the Upper Mississippi.

off the link to the backwaters. Ecologists despair over the lack of submerged aquatic vegetation in this reach, a missing plank at the base of the food web. In the refuge, at least, the upstream edge of each pool retains a mostly natural character. Backchannels snake through the surviving marshes.

As you head downstream in each pool, though, the water begins to stack up against the dam, spreading until it reaches from bluff to bluff, as I saw at Trempealeau. This diminishes fish habitat, since some species need flowing water. The slowed water drops its mud, reducing the depth of the backwaters—which increases the risk of oxygen loss and eliminates pockets of warmer water that help fish survive through winter. And the pools have swallowed the surrounding topography: sandbars smothered, forests inundated, old islands vanished beneath the water—240 square miles gone, as much land as all of Chicago.

The morning after my stay in Trempealeau, I met Tim Yager, the deputy manager of the Upper Mississippi refuge, on the edge of Pool No. 8. Old maps, created by the Corps in the 1890s, show what the river once looked like here: a forking channel, its two streams evocatively labeled as Raft Channel and Coon Slough, slip through land that is marked as marsh, which frays into its own tangle of smaller creeks and streams. On my visit, as I looked down from a small parking lot, there were no channels left, and no marsh. Just a watery expanse. Yager explained why that was a problem: the air rushes unhindered across the water, churning up waves. "With waves comes erosion," he said. "With erosion comes loss of islands. With loss of islands comes lower diversity in terms of habitat."

In 1986 the engineers began to rebuild the islands. Today, after several design improvements, they're shaped like crab claws, the pincers pointed downstream, an orientation that slows erosion. By reducing wind shear, the islands have helped underwater vegetation to re-emerge. Water funneled along the edges of the islands carves deeper pockets in the lake bottom, which have helped fish populations rebound.

Still, other stretches, even within the refuge, continue to suffer. The spreading water is slowly killing the silver maples that once dominated the floodplain; where the maples die, opportunistic new species like reed canarygrass and Japanese hops jump in, blocking the trees' ability to return. To add to the trouble, the amount of water pouring downriver has been increasing in recent years, causing more and longer flooding, killing even more trees. The result is an ecosystem so unstable that scientists aren't quite sure what changes might arrive in the decades to come.

The uproar in Alton in the 1970s was not just about fish. The National Environmental Policy Act had forced the Corps to open its books, which revealed how pricey river engineering could be—hundreds of millions of dollars, in the case of the proposal for the new lock. So this controversy prompted an economic reform, too. The Inland Waterways Trust Fund, established in 1978, is filled by a tax on fuel burned by commercial boats on the nation's rivers—twenty-nine cents per gallon, currently—and helps cover the costs of construction on new facilities. Still, since operation and upkeep of the locks and dams remain a federal responsibility, some estimates put the taxpayer burden at 90 percent or more of the system's costs.

Nor did this new payment structure do much to stop the torrent of misguided projects. In 1984, the Corps finished carving a 234-mile-long trench through Alabama and northern Mississippi, linking the Tennessee River to the Tombigbee, thereby providing an alternate route south to the Gulf of Mexico. The system required ten locks and cost $2 billion. The Corps of Engineers predicted 29 million tons of cargo would pass through the canal annually, but the actual flow has hovered around a third of that amount. Perhaps the world didn't need a second, artificial waterway running mostly parallel to the Mississippi River, 150 miles to the east.

Even on the Ohio River, always a heavily trafficked waterway, the Corps of Engineers has stumbled into boondoggles. Olmsted Lock and Dam, near the river's mouth, was meant to replace two of the oldest remaining dams. The installation was supposed to take seven years and cost $775 million. Instead, due to an attempt to use innovative new construction methods, the project turned into a thirty-year, $3 billion ordeal. The Inland Waterways Trust Fund grew badly overdrawn. Originally, the industry had covered half the cost of any new lock-and-dam construction. In 2020, to solve the problem, Congress reduced this responsibility to 35 percent.

Defenders of the navigation system point out that it shaves $7 billion to $9 billion off shipping costs each year. But who benefits from those savings? More than half the cargo on our inland system consists of coal and petroleum—so this amounts to a massive subsidy for one of the nation's dirtiest sectors. Farm goods, which feel slightly more wholesome, make up a substantial portion of the remaining cargo—but most of the grain shipped on the river is meant for export, not for domestic consumption. (And don't think of grain as being food, necessarily: much of it is broken down to create, among other things, biofuels and bioplastics, fish feed, rubbers, fibers, solvents, and adhesives.) Only certain cargo can be shipped by barge—the boats' slow plod is a challenge for anything perishable—so railroads can raise rates on other products to make up the difference. Officials like to declare that, as an army general once put it, the "well-being of the Midwest" depends on flowing towboats. But this is a vague proclamation, unsupported by much in the way of firm numbers. "There is a vast gap in the economic literature regarding the importance of these projects to regional economies," Donald Sweeney, a former Corps of Engineers economist, told me.

Sweeney spent five years in the 1990s analyzing the costs of further lock expansion projects along the Upper Mississippi. Ultimately, he determined that the delays the boats endured were not enough to justify the expense of replacing the locks. But when he shared these results with his army bosses in 1998, he was pushed aside and replaced by a new team. Sweeney was still cc'd on the flow of emails, which is how he saw that his replacements had received orders from an army general "to develop evidence or data" that could justify the expansion project. As he watched the data turn from fact to fiction, Sweeney decided to inform the Office of Special Counsel, a federal agency meant to investigate governmental malfeasance. He also leaked documents to Michael Grunwald, a journalist at *The Washington Post*, who went on to publish a series of stories criticizing the Corps's wasteful behaviors. The documents Grunwald uncovered revealed that the Corps of Engineers was pursuing what it called "Program Growth Initiative," an effort to find more projects.

One can see the rationale: the more projects the engineers recommended to Congress, the more money Congress was liable to send back their way to build them. Legislators, meanwhile, are happy to tuck such infrastructure into massive spending bills, which are too dense for journalists to parse, because later they get to stand beside a big new lock and brag about how they helped make it happen. "People like you or I or my mother, who pay the taxes and actually foot the bill for the project, get no say in the process whatsoever," Sweeney said. The brazenness of the scheme was shocking, even comical; one of Grunwald's editors joked that the resulting series of stories felt like coverage of communist Czechoslovakia. The Office of Special Counsel eventually noted "serious misconduct and improprieties" within the Corps of Engineers and such a bias toward securing big projects that the agency struggled to maintain objectivity in its analysis.

As time passed, the scandal settled, and the Corps of Engineers ran the numbers again. This time they admitted that it's difficult to predict the future of river traffic. That's why they presented a range of future scenarios—what might happen if traffic grew, but also what might happen if, despite expanding the locks, traffic lagged. "What the Corps was good at doing was obfuscating," Sweeney told me, summarizing the new results. " 'It could be as low as this; it could be as high as that—we don't really know.' The range was so wide that, whatever you believed, you could find a number in there." In the end, given the possibility that traffic *might* grow—at what Grunwald called "outlandishly optimistic" rates—the agency recommended an even more expensive proposal than the one Sweeney had rejected. To make this suggestion more palatable, the engineers paired the proposed navigation improvements with a promise to spend "comparable" dollars on ecosystem restoration. The wording was so vague as to be meaningless, but big conservation groups like the Nature Conservancy signaled their support—eating up the "green pork," as Grunwald wrote. The new project became known as the Navigation and Ecosystem Sustainability Project and was approved by Congress in 2007. Now, piece by piece, it's slowly being built.

Lately, the ecosystem dollars have targeted the rivers' most hated villain: Asian carp. This ecological problem can't be pinned on the navigation industry. These fish were first brought to the United States in the 1960s as, ironically, an ecologically friendly way to clean the fishponds that were then emerging in the farmland along the lower river. Eventually, they were released from laboratories into the local sloughs and bayous, which offered a link to the Mississippi.

Asian carp describes four species, but they've become infamous in large part due to the strange behavior of silver carp—which, when startled by the rumble of a boat's engine, fling their great heft out of the water, sometimes knocking boaters and water-skiers unconscious. The real danger of these species, though, lies in other facts of their biology. They're incredibly fecund, laying up to a million eggs at a time; they reach maturity in as little as two years and live for twenty; and they have few predators. One Corps of Engineers report to Congress offers what it declares to be a "fun fact!": "Bighead and silver carp lack a true stomach, requiring them to eat almost continuously." Since their preferred food sits at the bottom of the food chain, this endless hunger wreaks ecological havoc. There are places on the Upper Mississippi where Asian carp now make up nearly two-thirds of the fish community. The biomass of native fishes, meanwhile, is plunging.

The war against Asian carp offers a tribute to the human imagination and all its dark possibilities. We've dumped poisons into tributary rivers. We've built computer models to track three-dimensional river flow, trying to understand how carp eggs are distributed so they can be intercepted and removed. Companies have popped up to buy the carp from fishermen; once filleted, the fish is flash-frozen and shipped to Asia or chopped up and molded into fish cakes and dog food. Portions of Pool No. 8 were closed a few months before my visit; biologists riding in boats rigged with underwater speakers drove the carp into netted "cells" within the lake. Most of the carp were removed. Five were outfitted with acoustic transmitters to track their subsequent travels.

Despite such efforts, these fish are unlikely to be eradicated. The best we can do at this point is stop their further spread. That's why in 2010 the water in a Chicago canal that connects the Mississippi watershed to Lake Michigan was electrified: the current is meant to kill any invaders brave enough to try to enter the lake and all the many waterways beyond. Then in 2015, the Corps of Engineers announced its plan to shutter the river's northernmost lock, at Upper St. Anthony Falls, in Minneapolis. The lock once allowed boats to bypass the river's sole waterfall—but gave the hated carp a potential path northward, into the delicate headwaters ecosystem.

Though fears over the invading carp were prominently featured in headlines in 2015, they were not the sole prompt for the lock's closure. They were not even the main cause. This lock had always been one of the system's great boondoggles. Its forty-nine-foot ascent—to a city-owned harbor upstream of the waterfall—was a trek few boats bothered to make. Alongside the waterfall, the Corps could build a lock only big enough to fit two barges at a time, and there are plenty of easier spots to unload downstream in St. Paul, or up the Minnesota River. When the money-losing harbor shuttered in 2014, the Upper St. Anthony Falls Lock and Dam was due for repairs, and the cost was hard to justify. So Congress decided to close the facility.

Minneapolis straddles geology rare for the Mississippi: the river runs through a narrow ten-mile canyon. The rocky rapids at the canyon's bottom are gone, buried under the water held by the city's two other dams—but these dams are now mostly useless, too, along with their accompanying locks, because there's nowhere else in the city to unload.

Several groups in Minneapolis would like to close the locks and rip out the dams. That would almost immediately bring the rapids back to the city—turning Minneapolis into a potential urban paddling mecca.

Officials from the towboat industry, meanwhile, have expressed concerns that the sediment that's trapped behind the dam may contain toxins that could be flushed downstream. They've emphasized Congress's careful language, too: the lawmakers made sure to declare that the closure at St. Anthony Falls was "unique" and not precedent-setting. Someone managed to slip in a caveat that "tonnage is an arbitrary metric"—as if the lack of cargo passing through the lock were irrelevant. Sometimes it feels like the industry wants to ensure there are no steps back: they'll find a reason to keep even the most useless infrastructure. Perhaps they worry that if even one dam comes out, people might like the resulting river. That rip in the canvas, that glance at the original river, might give people a chance to envision a future when the whole of the old Mississippi might return.

When I visited the refuge in the Driftless, I could not help but wonder what such a change might mean for this strip of river. Yager, the deputy ranger, was not enthusiastic. He noted that many lakes and wetlands have been drained across the surrounding countryside. From the point of view of a canvasback duck, say, having a bit of open water along the Mississippi River is a boon. If for some reason navigation ceased along the river, Yager suggested the dams could be used as water control structures. Sometimes the flow might be reduced, exposing more mudflats so that vegetation could reestablish itself, restoring some diversity, before the pool was filled again. (Indeed, this strategy is already conducted occasionally, though at a scale small enough to ensure the navigation channel remains unchanged.) That would be good for the ducks, but not so much for the fish who prefer a long and unbroken river. So how do we decide who wins? In 2018, when the government tasked a team of scientists to determine what might be considered a healthy Mississippi, they essentially tossed up their hands. Given all the changes on and off the river, and given all the competing visions of what a river should be—its "several purposes," as Congress put it—they knew they'd never be able to come to a consensus about what sort of ecosystem should be our goal.

~~~

To the west, by contrast, on the Missouri, the Corps of Engineers once had a clear ecosystem goal: it had to build a river that could suit the pallid sturgeon. Its hands were tied, really, thanks to a legal ruling in 2004. That year, after a decade and a half of courtroom wrangling, a judge declared that the Corps had lapsed its duty to protect this endangered species—and had thirty days to write up a new plan. Before the year was out, backhoes began to cut notches in more than five hundred dikes, removing rock from the middle to reopen back-channel flow. The Corps removed some dikes entirely. The engineers cut new channels through the mud. They began to spread and slow the narrow river that had been created a half-century earlier, for the sake of towboats. The sturgeon's old universe was being returned.

Then a few years later came another lawsuit: farmers along the river blamed all this work for a sudden increase in floods. Throughout the next decade, the plaintiffs won victory after victory after victory, and in 2023 a federal appeals judge affirmed the earlier rulings: the restoration program, by causing more flooding, constituted an illegal taking of the farmers' land. (The "interception-rearing complexes" I tried and failed to see on the Missouri were a smaller-scale intervention launched amid the legal struggles; even these have proved unacceptable to the farmers and the towboats—who are emboldened, perhaps, by their wins in court.) The ruling may be consequential for the whole watershed. The judge suggested that when the engineers turned the Missouri River narrow and navigable, they implicitly promised that this was how it would be forever. That, to me, sets an alarming precedent. We're bound to decisions that our predecessors made back in an era when, even had anyone cared to think what a sturgeon might want, we lacked the science to know. We're committed to an infrastructural approach that considers rivers an enemy that can be conquered.

Against this backdrop, the Lower Mississippi stands apart: there is
~~~

little furor. Nearly a third of the eight hundred dikes on the lower river feature the kinds of notches that sparked the lawsuit on the Missouri, yet these have prompted no legal complaints from the agriculture and navigation industries. The ecological benefits of the notches, meanwhile, have become undeniable. The resulting sandbars have been the salvation of the least interior tern, a bird that was listed as endangered in 1985, but whose population has jumped tenfold. The species is now so booming now that, as of 2021, it has been officially "delisted" across its range. The latest step in the river's revival is the addition of wood traps in the backchannels, structures meant to retain some of the driftwood floating downstream—a return, if only in small pockets, of the snag habitat that Henry Shreve removed two centuries ago. The biologists expect these new tangles of wood will become recreational hotspots, given the desirable fish species they're likely to attract. The U.S. Fish and Wildlife Service, after reviewing all these efforts, has declared that so long as the Corps of Engineers maintains connections to eighty-four backchannels along the lower river, the Mississippi River and Tributaries Project poses no threat to any known endangered species, including the pallid sturgeon.

Such is the power of a big river. While 90 percent of the lower river's floodplain is gone, severed by the levees, the 10 percent that remains constitutes two million acres. This mass of wetlands, while not officially protected, is an order of magnitude larger than the Upper Mississippi River National Wildlife and Fish Refuge. It's a strip of forest that runs for a thousand miles along the river, stretching as much as fifteen miles across. Ironically, the batture is widest where Harley Ferguson snipped the river's bends: the loops of river he left behind were already contained inside the levee system, and no one wanted to move the great wall. So unlike the Missouri River, where farmers have pressed as close to the river as they can get, the Lower Mississippi has room enough for boats and birds and fish.

Because there are no dams below St. Louis, the Lower Mississippi retains a more natural hydrograph, too, rising and falling as much as thirty feet in a typical year, and twice that in a big flood year. This movement creates an ever-changing tapestry of bars and forests and inlets, landforms that are alternately smothered and exposed. Oxbows are connected by a pulse of water that is like a heartbeat, keeping the river's ecosystem alive. Fish slip into these quiet waters to feed on the algae and spawn and rear. The vegetation pulls fertilizer out of the river, too, reducing pollution downstream.

Dirt roads run throughout the batture, servicing timber farms and hunting camps, but there's almost no pavement. The entire state of Mississippi, which features four hundred of the river's wildest miles, offers only four highway crossings. Really, the only way to see this place is by boat. From a canoe, you can note raccoons prowling the shoreline, deer ducking into the woods, a ghostly presence that may be a coyote but disappears before you can be sure. You can camp on empty islands that tower like cliffs, dozens of feet above the water, when the river is low. One of my island campsites, near Hickman, Kentucky, did not appear on the navigation maps I carried. The maps had been published ten years earlier, which establishes an upper bound for the age of this bit of land. By my estimate, the island already spanned more than a mile in perimeter, all covered in a willow forest—new habitat for all the wildlife. Nearly two hundred such islands have been born over the past sixty years, an unintended consequence of the wing dams, which trap the river's sand. Though the endless flow of sediment causes problems, too: many of the old oxbows are becoming plugged with this mud, cutting off essential habitat. Abandoned oxbows have always filled, but in earlier eras, the river's meandering would carve new backwaters. Now, with the river trapped in place, no new meanders appear.

I swim in the river on these trips. I drink several cups of river-water coffee each morning. There are places where caution is worthy—in Louisiana, especially, where industrial plants line the riverbanks—but thanks to environmental laws of the 1970s, the Clean Water Act especially, many stretches of the Mississippi and its tributaries can be held up as examples of the possibilities for recovery. Over the subsequent decades, the number of bacteria in the lower river plunged, as did amounts of lead. Not that the new laws were a panacea. Solid waste is not covered; the Gulf of

Mexico now holds among the highest concentrations of plastics of any ocean in the world. I see the refuse when I paddle the river, washed up onto islands: Coke bottles, shopping bags, hard hats. Once, disturbingly, I came across a plastic baby doll, eyes smudged with dirt, lying like a corpse amid flood-carried driftwood. One morning in New Orleans I walked to the riverbank and found the sand covered in nurdles. These tiny, pebble-size globules—the synthetic "raw material" that is melted down in factories and reshaped into anything plastic—had fallen off a cargo ship. In the weeks that followed, no one was punished for the spill, since the pellets are not classified as a pollutant.* The clean-up had to be conducted by volunteers.

Even tarnished, the batture is "one of the most important remaining wilderness areas in the United States," as the Corps of Engineers put it in a 2018 scientific report. One of the authors of that report, the biologist Jack Kilgore, told me he figured the batture along the Lower Mississippi was one of the world's most productive floodplains—perhaps beat out only by the Amazon. This is one of the strange truths about the Mississippi River: despite centuries of abuse, its southern miles remain a beautiful place—one that, hidden behind the levees, the world has mostly forgotten. For those who have spent time out there, the river remains the same alluring frontier that it was two hundred years ago, a place you can spend weeks camped on sandbars and see almost no one else. It's not the same river it always was, but then again, this river has never stayed the same.

None of this is to say that the Lower Mississippi River is a symbol of hope. Not every river is so big, so able to accommodate so many competing visions. There is something-old fashioned about this landscape. Once we feared the wild: we built fences to keep the wolves from our pastures—and then, when that was insufficient, did our best to wipe out the beasts. These days we carefully preserve and protect our favorite ecosystems,

* Nevertheless, when a pair of activists dropped a box full of nurdles on the doorstep of a chemical lobbyist, they were booked for terrorism. Thankfully, they were never formally charged.

and the fences are as much as about keeping the wildness *in*. The batture is tucked behind its wall because we cannot control the great river. So here is a last strip of wilderness in the old style, a land that America has never claimed. The batture, then, is less a sign of what is possible than a reminder of what is not. We humans have our limits.

The Dow Chemical Plant, as seen from the head of the Bonnet Carré Spillway (St. Charles Parish, Louisiana).

10

Death Alley

The precarious life on the levees' dry side

The world we've built atop the stolen floodplain, on the dry side of the levee, can be a confounding place. Its endless sameness—"nothing but flatness," as Eudora Welty once put it—turns this land into an inverted labyrinth, perplexing not for its abundant corners and pathways but for its lack of any waypoints at all.

The first time I visited the Little River Drainage District, as I sat in the passenger seat of an oversize pickup truck that barreled down the unswerving farm roads, I felt a sense of dislocation that doubled as déjà vu. The scenery was undeniably familiar, but it seemed reshuffled, as if someone had pulled an elaborate prank and rearranged all the street signs. Each acre of farmland looked like the farmland I'd been seeing for years—in Louisiana, in Arkansas, in Illinois. The only distinction now was the gloom of winter. The turnrows were empty, glazed in a wet gray film of rain that matched the wet gray smear of sky.

My tour guide, W. Dustin Boatwright, was the chief engineer of this drainage district, the country's largest, which steers the rainfall from a half-million acres in Missouri's bootheel off the farmland and into a network of ditches and canals. Boatwright's grandfather moved to Missouri in the 1950s—he had been "making his way up the Delta just chasing work," Boatwright told me—and found a job as an oiler, keeping the excavators used by the drainage district ticking. Soon he was leading field crews, work he passed down to his son. Boatwright figured he'd keep up this family tradition, but his elders pushed him into college. Which is how in 2014, as a twenty-nine-year-old, he became the youngest chief engineer in the district's history.

Despite the office job, Boatright seemed to retain the ruddy health of a man who worked with his hands. He'd greeted me that morning at an office park in Cape Girardeau with a firm handshake. His style lacked any concerted flash—he wore a run-of-the-mill button-down tucked into khakis—and in good Midwestern fashion, he seemed determined to address me as "sir." As we drove, he told me his favorite president was Teddy Roosevelt, in part because Teddy, much like Boatwright himself, wasn't a party guy; he had forged his own path through the politics. But Teddy also struck a balance when it came to nature. He preserved some bits while still spurring the nation forward; he loved the big forests but knew some had to be sacrificed for the nation to grow. We could see some of that sacrifice through the windshield. "Everything around where we're driving right now was swamp," Boatwright said, gesturing toward the monotonous farmland.

I noted that after studying the river's past, I'd come to think maybe we might want some of that swamp back. Boatwright surprised me: he turned a bit prickly. Did I think the government should just come in and take people's land? he asked me. When it comes to restoring wetlands, he observed, there are important things to consider. "Any area that's left untouched or wild is ever-changing," he said. "Now, when you start—"

Then he realized that one of the ditches he'd meant to show me was whizzing past the window. "I'm sorry, I'm terrible," he said. And with that, as if we'd never broached the difficult politics of this landscape, he began to explain the intricate system of canals.

As Boatwright drove south across his domain, I watched the "laterals" pass outside the window. The canals crossed the Little River Drainage District in east-west lines and provided a steady backbeat to the

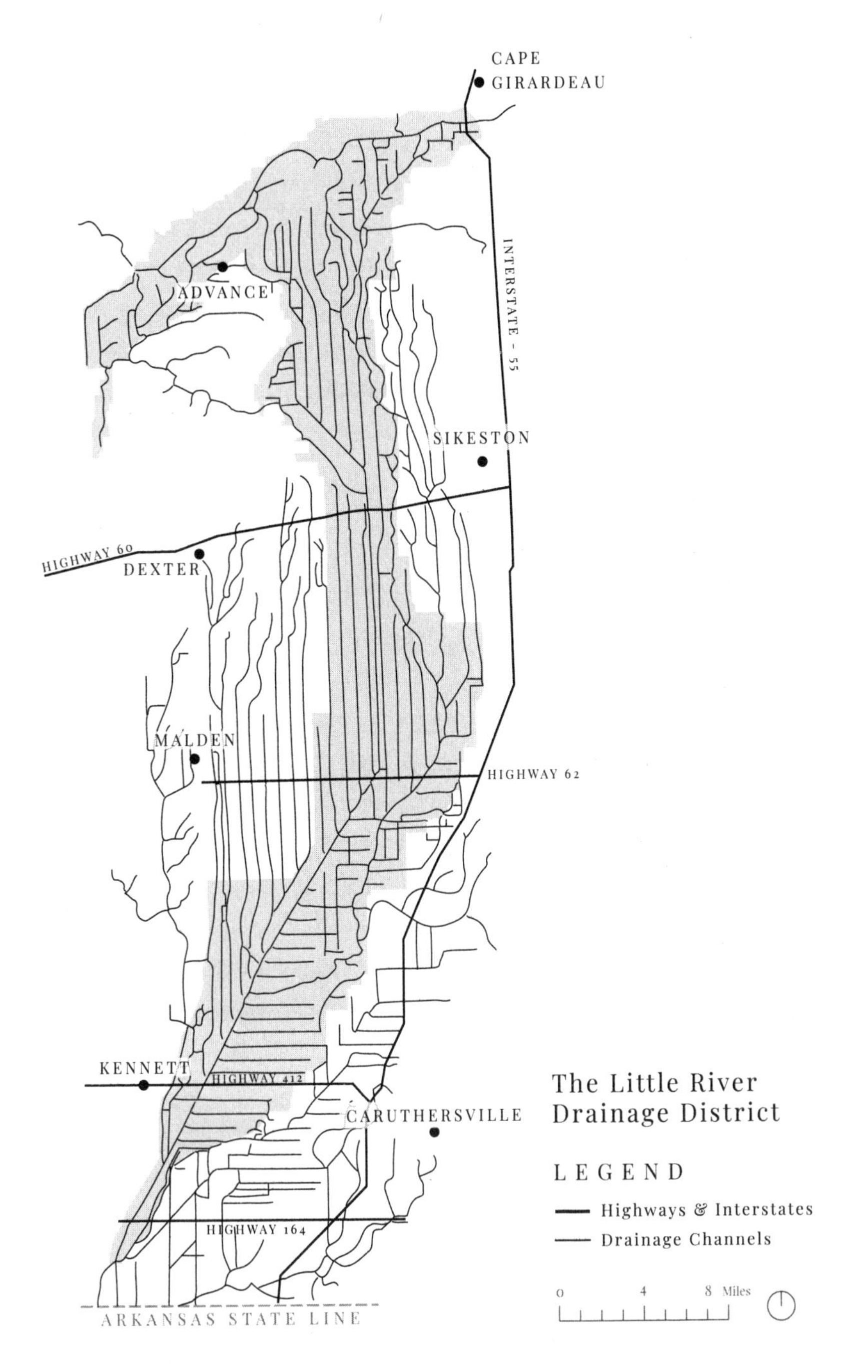

The network of canals managed by the Little River Drainage District drains a half-million acres in southeastern Missouri.

landscape—a line of blue every mile or so, splitting the constant blurred gray-brown of the wintertime soils. The laterals, standing ten feet wide and ten feet deep, come together in a set of five southward-running outlet channels. They are the district's major drainpipes—a human-made replacement for the Little River, the sluggish, low-lying waterway that once mired this region. The widest outlet channel spans 150 feet at its bottom, which would make this a good-size river were it not so straight, so obviously an artifact of humanity. In total, the district includes nearly a thousand miles of ditches and more than three hundred miles of levees. Boatwright estimated that when the contractors dug these waterways in the early twentieth century, they moved more dirt than was hauled out of the Panama Canal.

Similar canals run throughout the whole of the alluvial valley. So many swamps have been drained and cleared that the lower river is visible from the International Space Station. Most of the South appears as a swath of trees, where the forests, though thinned by farms and cities, remain robust enough to leave a fuzzy green texture. Through the center of the country runs a mottled brown band. This is the alluvial valley. The color is the result of the old floodplain's near-absolute treelessness, its conversion to an endless empire of dirt and crops.

As the trees have gone, we've lost so much. In 1821 John James Audubon watched hunters kill tens of thousands of golden plovers on the outskirts of New Orleans in just a few hours. Today, you'd be lucky to spot more than a dozen plovers in a day. The cougar and the red wolf have been driven out of the river valley entirely. The last undisputed sighting of an ivory-billed woodpecker—whose blaze of red feathers and great size spawned so many exclamations that it earned the nickname "Lord God Bird"—occurred in 1944, in the Louisiana bottomlands. Soon afterward, despite the objections of four governors, the land where the bird was seen was sold to a timber company. Every few years an ornithologist announces that they've found the bird again, but the claims always fizzle under scrutiny. The photos are too grainy, the bird's image impossible to distinguish from a look-alike species. It's as if we're haunted by what we've killed.

Boatwright, an avid hunter, said he'd have loved to see those forests. But he also emphasized that the old ecosystem here was always changing. Now, sure, humans had changed it again—but we'd gained a lot along the way. In a rough estimate, he put the value of land along the major local tributary, the St. Francis River, at over $12 billion. From this point of view, the brown expanse of the alluvial valley represents a staggering accumulation of riches.

Our conversation in the truck lurched back and forth between Boatwright's detailed elucidation of the landscape—its geological history; the crops that are grown here; the species of native grasses that the drainage district has planted on the banks of its ditches—and a larger clash of political philosophies.

"If you have a rural and an urban setting, would you say that that urban setting is dependent on the rural?" Boatwright asked at one point.

"Yep," I replied.

"Would you say the rural is dependent on the urban?"

"Yeah."

"Absolutely not," he said, like a good Jeffersonian, envisioning this landscape as a self-sufficient world.

In my view, the closest thing the alluvial valley ever had to Jeffersonian cultivators were the Choctaw farmers who were "removed" two hundred years ago. Throughout its American years, this place has produced very little food and has always depended on bringing money in and sending crops out. The first Mississippi River planters needed outside banks to fund their ventures and foreign companies to buy their cotton. Today the major crop is soy, but the program remains the same. Even more damning to this idea of self-sufficiency is the way government intervention has repeatedly saved this place. To see the modern floodplain clearly, we need to take a step back and consider another form of engineering: not just the physical interventions that drained and dried this place, but the economic policies that allowed it to survive.

Consider the Little River Drainage District itself. In 1920, just as the engineers were completing their initial work on the canals and levees, the price of crops began to tumble. Before the end of the decade, more than half the landowners in some parts of the bootheel were delinquent on taxes. Vast tracts owned by timber companies—tens of thousands of acres—lay idle, and their owners saw no need to pay taxes now that the trees were gone. The Little River Drainage District cut back on maintenance and still was unable to pay its bonds.

The falling crop prices turned out to be just an early warning of the coming Depression. By 1931, tens of thousands of tenant farmers in the alluvial valley were starving. Many had been forbidden by their landlords to plant anything but cotton, and now they had nothing to eat. The planters' prospects were sinking, too: the year before the market crashed, U.S. cotton farmers had earned $1.5 billion, but by 1932, their income had dropped by two-thirds. What saved this farmland was not some magical cultivator fortitude but President Franklin Roosevelt's legislative barrage.

The New Deal is famous today for its massive public works program. For farmers, though, and especially farmers in the alluvial valley, what mattered most was an overhaul of the industry's economic structures. Federal loans helped bail out the struggling drainage districts. New subsidy programs, meanwhile, propped up the farmers. Among other strategies, the government began to pay landowners to leave their land fallow, an attempt to dampen supply and thereby lift prices.

This was a national program, but since the nation's biggest farms were in the alluvial valley, the planters here swallowed huge chunks of the funds. A single farm in Mississippi County, Arkansas, just south of the Little River Drainage District, brought in $167,000 in payments in a single year (nearly $4 million in today's money). For the first time in long memory, many planters rose out of debt. "People come in here and ask to pay back interest on notes we literally have to fish out of the waste baskets," a banker in Tunica, Mississippi, reported. After the Depression ended, the payouts continued—growing, eventually, into the complex web of loans and payments and insurance programs that still undergirds the rural economy.

So just as the Mississippi River and Tributaries Project was transforming the river, this influx of money was changing life within its floodplain.

Before the Great Depression, only the wealthiest farmers had tinkered with tractors or dropped chemicals from crop dusters. By and large, the industry still depended on human labor—on tenants and sharecroppers. On paper, the government programs were supposed to sustain the U.S. farming tradition, keeping the flickering dream of Jefferson's empire of liberty alive. So some of the money was supposed to go to the tenants. But planters saw no reason to pass the payout on. They evicted the unneeded tenants. The money they kept could go toward tractors, so that once they began to plant again, they wouldn't need the labor. Government-funded agronomists, meanwhile, began to push the new technologies meant to increase farm efficiency—which only sped up the trend of consolidation and mechanization.

It was not hard to predict the effects of such programs. Indeed, in the 1950s, the Citizens' Council, a less-violent cousin of the Ku Klux Klan that was launched in the alluvial valley, explicitly cited tractors and cotton pickers as an effective means to drive hundreds of thousands of Black residents out of Mississippi. A decade later, once that mission was accomplished, thousands of old tenant homes were burned to clear space for more crops. Plumes of smoke rose into the sky, providing, in the words of one historian, "visible evidence that the revolution in agriculture had reached its final phase." In total, in the four decades after the Great Depression began, the farm population across the Lower Mississippi Valley fell by 78 percent.

For a brief moment, early in the twentieth century, smaller farmers were buying up lands in the western alluvial valley, Missouri included. But in the years after World War II, as the land-clearing continued, almost all the new land in the bootheel was tacked onto existing farms. As the farms grew larger, the local villages shriveled and shrank. Once, on a long canoe trip, I came ashore in New Madrid, a riverside town in the bootheel. The walk to the highway for supplies was dispiriting: the ad hoc home repairs and cracked storefront windows led one of my companions to remark that it felt like entering a foreign land—the kind of place that was once known as the third world, still straining toward the comforts of modern capitalism. But this town, I realized, wasn't headed anywhere. It had been jettisoned by our brutal agricultural economy.

The consolidation of so much money and power in so few hands did not go unchallenged. In 1934, after an Arkansas planter evicted a group of tenants who had just put in their cotton crop, eighteen men—some white, some Black—assembled in a nearby schoolhouse and launched the Southern Tenant Farmers' Union. Their rhetoric reconsidered the Jeffersonian ideals of property. "Since the earth is the common heritage of all," the union's constitution declared, "we maintain that the use and occupancy of land should be the sole title."

Their fight for higher and more equitable wages was met with violence. Some planters, proudly calling themselves "night riders," forcibly broke up meetings. Gunfire rattled through homes. Still, over the next few years tens of thousands of tenants across the country joined the union. At one point in 1939, evicted sharecroppers in the Missouri bootheel stacked their possessions along Highway 61; the protest garnered sympathetic coverage from national newspapers. But as the need for farmworkers waned, so did the union's influence.

The federal government made some efforts to improve the lives of tenants, too. A New Deal program launched rural colonies across the country, places where displaced sharecroppers might find a foothold and, if things went well, eventually accrue enough capital to buy the land they worked. More than a dozen such towns were built in the Arkansas portion of the alluvial valley alone. But critics quickly branded the resettlement sites as "un-American social experiments." Handouts to landowners were fine; helping more people become landowners was not acceptable. In 1943 Congress acceded to the wealthy men's wishes and axed the resettlement program, commanding that the colonies' assets be liquidated as soon as possible, sold off to private owners. The country singer Johnny Cash grew up in one such colony in Arkansas. His home, by luck, is now one of the two dozen that remain in the fields around Dyess: a trim white cottage that stands alone, incongruous, amid the fields. "They just tear them down to make more farmland," my tour guide told me when I visited.

The big farmers were supported by a powerful champion: a white man from the hills above the Yazoo Basin named Jamie Whitten, who in 1949 became the chair of the congressional subcommittee that distributed agricultural appropriations. He held the post for nearly fifty years. More than half of his district lived in poverty; in a typical year in the 1960s, they'd received, in total, $4 million in food relief. Whitten suggested that "giving people something for nothing" was more likely to destroy character than address nutrition. Nonetheless, that same year a group of planters who represented less than one percent of the district's population received $23.5 million in federal payouts to help support their farms.

Even Black farmers who owned their land suffered, since whites dominated the local commissions and committees that oversaw the distribution of government funds. Black farmers were given worse interest rates; they were asked for excessive collateral on loans; they were assessed for taxes at fraudulently high rates. This was nothing short of extortion: their budgets strained, Black landowners tumbled into foreclosure; the land was bought up and appended to the massive, white-owned, government-subsidized farms. Between 1950 to 1964 alone, Black farmers lost eight hundred thousand acres in Mississippi. A research team recently calculated the total cost of the loss, including lost income, was somewhere between $3.7 billion and $6.6 billion. Still, despite the losses, despite the great outmigration of Blacks, white people remained a minority in the alluvial valley. Through the decades, white families began to cluster in the larger and more prosperous towns, where they could hold more political power. The smallest villages, meanwhile, turned into nearly all-Black enclaves—places where poverty is rampant and the tax base is all but nonexistent. This has created a "new type of black ghetto," as one geographer wrote in 1990. The Yazoo Basin, he noted, was more segregated than it had been four decades earlier, before the Voting Rights Act.

Today big farmers in the alluvial valley possess tens of thousands of acres, millions of dollars' worth of land. Some of the biggest "farmers" are hedge funds and other institutional investors. The pension fund TIAA, for example, owns 130,000 acres across the alluvial valley. Through shell companies, Bill Gates, reportedly the largest agricultural landowner in the country, holds over 100,000 acres in Louisiana and Arkansas, including a single twenty-thousand-acre farm in the

Mississippi's former floodplain. The federal gifts still flow to these farmers. Over the past two decades, the government has paid out almost $1.5 billion in crop insurance along the Mississippi River. There's been talk lately of expanding this coverage into the Mississippi River batture, a scheme that clarifies how full of handouts our agricultural economy has become. The nominal goal of this move is to create more "productive" land, but the batture land is all but guaranteed to flood, so it won't be agriculturally productive, just productive in the sense that its owners will be sure to make money. Economic modeling suggests that little of the value of these subsidies ripples out into the rest of the rural economy, in part because after years of outmigration, there is so little rural economy left. Indeed, in much of the alluvial valley, the prospects are so paltry that a poor child's chances for economic advancement are worse than in any other developed country.

One could argue that now that this land has been cleared, we should keep farming most of it, and farming it as intensively as we can. The world's population will soon reach ten billion, and all these people need to eat. The more food we squeeze out of every acre, the less forest we'll have to clear elsewhere. But for this argument to make any sense, the alluvial valley would have to start growing actual food. Given the low margins on row crops, that might be good for farmers, too. In 2009 an Arkansas farmer named Shawn Peebles was so broke that he had to auction off all his equipment. The next year, inspired by a neighbor, he rented two hundred acres and began growing organic soybeans. "I made more money that year than I'd made my entire life," he told me when I visited his farm. Now he grows not just soy for edamame, but organic sweet potatoes and green beans and black-eyed peas. Peebles credits the shift with keeping his family's three-generation tradition of farming intact, but the change ripples beyond his family, into the community: to grow these vegetables, Peebles needs a crew of full-time employees. "We bring in fifty to sixty guys a year, into a town that has eight hundred people," he says. They're seasonal employees, arriving by visa, mostly from Mexico, but still their presence provides more of a jolt to the local economy than the typical row-crop farm.

To the south, in the delta, the details differ, but the general story has been the same: the ecosystems have been tarnished and the neighborhoods emptied, while a few big landowners have found a way to pocket real money.

The delta's version of this narrative has been shaped by the geological gift granted by the river's long presence here. Millions of years of river mud, laid atop the corpses of ancient life—the creatures that swam in this ocean and the plants that were carried downstream—has through heat and pressure produced an extensive collection of easily burnable hydrocarbons. Roughly 10 percent of the world's recoverable oil and gas lies in the Gulf of Mexico, a bounty that made headlines at the turn of the twentieth century, when prospectors unleashed huge gushers.

After the Great Depression, Queen Sugar never regained her throne. Just as the MR&T Project was being installed, farmers in the Midwest began growing sugar beets; industrial sugar production emerged overseas. But Emperor Oil came to the rescue. By 1937, the petroleum industry was already a lead employer in the delta, helping lift per capita income in some places to nearly the highest in Louisiana.

Oil does not burst from the Earth ready to burn; it must be refined. So new facilities appeared, first near the big cities of New Orleans and Baton Rouge. Then as the industry grew, the oil companies began to eye the rural parishes, where the old sugar farmland was cheap to buy and, conveniently, was protected by a brand-new system of levees. The state featured a friendly government, hungry for business. The final attraction—perhaps the most important attraction—was the river itself, which provided a convenient way to ship their product and, since it carried hundreds of billions of gallons of water, offered a good sink into which to dump and dilute toxic wastes. Companies like Marathon Oil bought up old plantations and then, sometimes unannounced and literally under the cover of darkness, demolished historic mansions so they could be replaced by coiling campuses of pipes and distillation tanks. Today, the cluster of ports between Baton Rouge and the river's mouth is the largest by volume in the United States. They service the more than two hundred chemical production facilities that line the river here. Just this narrow sliver of river mud—5 percent of Louisiana's landmass—can produce half of Louisiana's carbon emissions. Lately, as there is talk about shifting away from fossil fuels, the industry has pushed toward a new line of business: facilities that convert petroleum into plastics. (A similar buildout is occurring far upstream, along the Ohio River, another key reserve of fossil fuels.)

John Ruskey, the guide who first introduced me to the Mississippi River, once asked me to help him lead two photographers on a trip upstream from New Orleans. They wanted to capture the ethereal glow of the plants at night; the river, they knew, offered the best vantage. On the long days in our camps, on islands and sandbars, I lay in a hammock strung between willow trees, reading a book, enjoying the sunshine. Then after dark, we loaded into the canoe. Paddling upriver in the sluggish water just along the banks, we passed beneath dripping water-intake pipes and the trussed steel of barge docks. The plants loomed above the levees, skylines of coiled piping—"fractional distillation columns," officially. They looked like secret cities, protected by the moat of the river, the rampart of the levee. They were heart of an empire, then: America's Mordor, down to the flaming towers. Sitting on the water, under a night sky dulled into blue by the light pollution, I felt as if we were poised on the edges of a battle, ready to breach the defenses and pull the plug, sending a whole way of life shuttering into darkness.

Our drive back from that trip, into New Orleans, was disconcerting. Every drive along the river in the delta is. In the distance, flames forever rise above the levee, flickering from the towering waste vents. St. John the Baptist Parish is home to the country's only major aluminum oxide plant, which for every ton of product spits out a second ton of so-called red mud, a caustic radioactive dust that you can see smeared along the side of the highway for a mile surrounding the facility. Then there are the fertilizer plants, an ironic site here, in what is supposed to be one of the world's most fertile valleys. They're a signal of how far we've drifted from the region's natural ecology. In St. James Parish, a pile of phosphogypsum powder, a mildly radioactive waste by-product of these plants, now reaches two hundred feet tall, high enough to be noted on federal aviation maps. As you approach it on the highway, its bulk dominates the horizon, though you cannot see the hundreds of millions of gallons of wastewater that sit at the center, putting pressure on the surrounding pile. In 2018 one of its containing walls shifted more than a foot, forcing the plant's owner, Mosaic, to scramble to keep the water from bursting into nearby bayous.

So much of the landscape here was created by Black people: they cut the trees and chopped the weeds and planted the sugar and cotton. They built the mansions and barns and levees. They learned just how much pressure to apply to the plow, the right way to hold the saw, how to coax the mule. As the historian Walter Johnson points out, it was all labor, a part of an economic system, but it was also *work*, "the application of human energy and imagination to the physical world." Even today, a sense of connection remains: the land here is more than just land. Joy Banner, an activist who runs the nonprofit Descendants Project in St. John the Baptist Parish, discussed with me how Black families tended to bury their loved ones in the local soils. White planters, in contrast, often chose to have their bodies interred in New Orleans's formal cemeteries. Burial, as Banner noted, is a part of the agroecological cycle: as a body breaks apart, its chemical components add fertility to the soil. "The land holds the energy of those relationships," Banner told me. "It holds the bodies of our ancestors. Our connection with that land is just something that's more ethereal and powerful and spiritual."

Even before Emancipation, free Blacks had managed to obtain tiny slivers of land between the sugar fields—sometimes owning, sometimes renting, sometimes just squatting. Enslaved residents were permitted small plots where they could raise vegetables and chickens. After the Civil War, some of these plots were leased to Blacks by the Freedmen's Bureau or begrudgingly turned over by their white owners. This created a unique pattern of land use: huge properties, the kind that were tempting to industrial developers, sat next to smaller Black-owned plots that the banks and investors ignored. Throughout the twentieth century, the bucolic life on those small plots—where families grew butter beans and sweet potatoes, raised cows and pigs, and ate shrimp they'd pulled from the river—was replaced by something dingier. Almost everyone who lives

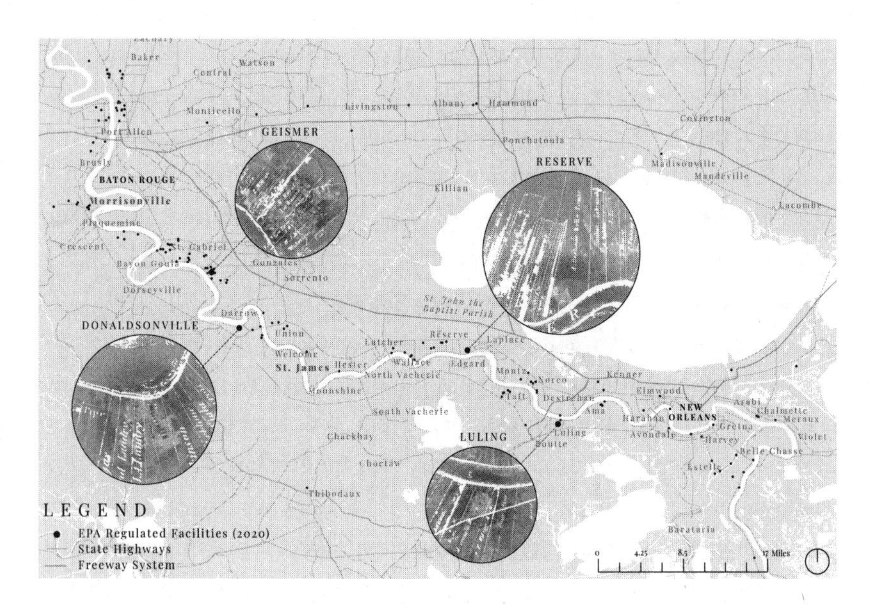

Between Baton Rouge and New Orleans, the Mississippi River features hundreds of industrial facilities.

here has a story to tell about cancer, whether it's struck their own body or that of a loved one. It's a signal of the invisible creep of toxins in the water and air and soil.

Sometimes the invasion was less creeping and more catastrophic. In 1960 a leaky refinery in Baton Rouge sent so many phenols into the Mississippi River that downstream residents could taste the chemical in their drinking water. In 1965 a towboat carrying six hundred tons of chlorine was caught in Hurricane Betsy and sank; as a precaution, the federal government shipped more than 110,000 gas masks to Baton Rouge. In the spring of 2012, Bayou Corne, which winds through the swamps to the west of the Mississippi, suddenly began to bubble. Then sinkholes appeared in nearby backyards. That August a pit the size of a tennis court cracked open, emitting a stink of diesel. It grew over several years until it spanned forty acres. Authorities eventually determined that the collapse had been caused by the reckless mining of nearby salt domes, in pursuit of industrial feedstocks. Because methane might have leaked out of the sinkhole, a small spark from flipping a light switch had the potential to set a house ablaze. Today the community is a curlicue of potholed roadways weaving between empty lots.

Back in the 1980s, amid a dispute over wages, a union put up a billboard along a nearby interstate that gave this stretch of riverland a grim nickname: "Welcome to Cancer Alley." Lately, activists have chosen an even grimmer name: "Death Alley."

One response to the specter of death and disease has been to move people out. The most famous example came in the late 1980s, when, after leaked toxins tainted the local water, the Dow Chemical Company decided to preemptively buy out an entire village. Morrisonville had been established by a group of formerly enslaved Black families in the 1870s; now Dow offered its residents as much as $200,000 to pick up and move on. Today all that's left of Morrisonville are its two old cemeteries, plus a few concrete slabs that once underlay the houses. These remains are inaccessible behind the fenceline of Dow's chemical plant.

Many local activists condemned the move: it was cynical, they said—addressing the symptoms but not the disease. Which proved correct. As the disease progressed for several more decades, more and more production facilities sprouted atop the riverbanks.

Lately, though, the industrial developers seem to have met their match. A burst of new community groups, like the Descendants Project, have refused to cede any further ground. I met Sharon Lavigne, an activist who has been instrumental in these efforts, at her house in St. James Parish, an hour upstream of New Orleans, on a tidy property that runs up against the levee. She told me that her activism is divinely inspired: in a moment of crisis, when she wondered if she should abandon this place, she sat down to pray and heard from God that she was meant to stay. So she launched RISE St. James. The nonprofit, partnering with other local and national organizations, has challenged a Taiwanese company's plans to build a massive plastics plant near her house—alleging that, among other problems, sloppy archaeological work had overlooked the presence of cemeteries on the property that held the remains of enslaved laborers. When a state judge revoked the permit for the plant, it was for a simpler reason: its emissions would be a threat to public health. "These are sacred lands," the judge wrote in her decision, quoting Lavigne's descriptions of how for generations Black families—enslaved and free—worked to keep the land productive and intact. Unfortunately, the permit was reinstated on appeal.

Not everyone wants to stay. In 2011 the nonprofit Louisiana Environmental Action Network (LEAN) conducted a survey and found that in one neighborhood near Sharon Lavigne's home, all but two of the 120 residents were ready to sell their homes and move on. Evenings here can be a cacophony—train cars coupling, river barges being loaded beyond the levee. Wilma Subra, a chemist who for years has been working on behalf of low-income communities in Louisiana, has detected unlawful levels of emissions in the air. One of the residents once suffered something like a "bad sunburn," as she described it, after she was caught outside in the rain. LEAN's founder, Marylee Orr, had been among the most vocal critics of the Morrisonville buyout. Now, two decades later, she felt that St. James Parish has grown so polluted that it was time to get people out.

Even when they're voluntary, buyouts are difficult. Some residents live in homes that were passed down through generations with little paperwork. The presence of so many industrial plants and pollutants depresses home values, which means even people with clear titles can't make enough money selling their property to buy a home in a safer neighborhood. At this point, without some kind of assistance, many residents are all but trapped. Michael Orr, Marylee's son and LEAN's communications director, suggests that for the locals who want to leave, the best approach may not be to stop the petrochemical buildout. Rather, they may have to let the companies come—on the condition that they cover the cost of the current residents moving out. St. James residents were seeking to leave their homes before the Taiwanese plastics plant was ever proposed, "and by my assessment, they need the opportunity to be able to leave," Michael said. "It's not a safe place to live."

The land to the north, in the alluvial valley, can seem bucolic in comparison—beautiful, even, in its mesmerizing expanse of green order. But it too has its toxic underside. As the agrochemical industry arose in the mid-twentieth century, the consequences here, where intensive agriculture was so dominant, were horrifying. During the 1957 spraying season, a doctor in Glen Allan, Mississippi, found more than a hundred patients outside her office every day. The men lay dazed on the grass, burning with fever, coughing up blood. Eventually the doctor herself grew so confused and muddled by the toxins that she had to spend a month in a hospital in Jackson. She tried to return to the alluvial valley, but once she drove north of Vicksburg, the chemical haze overwhelmed her. She turned around and never reopened her practice.

The wave of environmental legislation in the 1960s and '70s forbade the worst of these poisons. But fertilizers are exempted from the Clean Water Act, the 1972 law that oversees pollution in rivers. Between 1960 and 2000, an era when other pollutants were declining, the amount of nitrogen reaching the Gulf of Mexico increased as much as threefold. Nitrogen is considered a "nutrient" for crops, but in the ocean it feeds algae. When algae die, their cells decompose, a process that consumes oxygen. Each spring, as a new batch of fertilizer arrives, oxygen levels in much of the Gulf of Mexico drop to nearly zero, then stay low through the fall. The resulting "dead zone" is invisible from the surface, but if you dive down to the ocean bottom, you'll find an apocalyptic landscape of dead crabs and mollusks and sea worms, choked by the barren water. When the Gulf of Mexico dead zone was first measured in 1985, it spanned nearly four thousand square miles, larger than the state of Delaware. It's only grown through the years; at its most recent peak, in 2017, the dead zone was twice that size. Smaller algal blooms plague many lakes within the Mississippi system, too, releasing toxins that are hazardous for swimmers. Blooms routinely strike rivers throughout Iowa, which contributes more fertilizer runoff than any other state in the watershed. Local towns that depend on the rivers for drinking water are spending millions of dollars to clean their supply.

Herbicides, too—chemicals designed to kill weeds—have become a pressing problem. In the early years of the chemical revolution, herbicides were used sparingly, mostly in the off-season, since a poison that kills a weed is likely to kill the crops, too. Then in 1996 the agribusiness giant Monsanto released seeds that were genetically engineered to resist the herbicide glyphosate. Farmers describe that growing season as a miracle: it was as if the age-old problem of weeds had been solved.

The story of these magic seeds begins on the Mississippi River. Monsanto had long produced glyphosate—which it markets as Roundup—at a facility in Luling, Louisiana, amid the cluster of chemical plants. The company's scientists realized that the bacteria growing in pools of wastewater outside the plant had, through natural selection, developed an ability to withstand the chemical's deadly effects. They isolated these genes and inserted them into cotton and soybeans and corn.

Though glyphosate is considered safe by the EPA, the chemical has been implicated as a cause of cancer in several lawsuits. Bayer, the company that bought Monsanto in 2018, has paid out billions in settlements. That has not stopped the use of this herbicide. In the twenty years after the glyphosate-resistant crops were introduced, use of the chemical increased fifteen-fold. Indeed, it's been applied so much that, like the wastewater bacteria, some weeds have begun to evolve genetic resistance to the toxin. Lately, the seed companies have turned to new herbicides. Unfortunately, the next product, dicamba, has proved even more disastrous. And here the story loops back to the Mississippi River, because no region has seen more destruction from dicamba than the alluvial valley. In the 2017 growing season alone, the Arkansas plant board investigated nearly one thousand farmer complaints about damage from the chemical, spanning nine hundred thousand acres of farmland. Over the next few years, trained volunteers from Audubon Arkansas catalogued dicamba damage in local forests—in sycamore, oak, and pawpaw trees, among many others—across twenty counties, including city parks, cemeteries, and a wildlife refuge that is home to an endangered woodpecker.

Dicamba is not just one more assault on the old swamp forests. This chemical reveals the precarious state of the farmland economy we've built in their place. One of the most tragic reports I heard was about a small vegetable farm in Mississippi County, Arkansas, focused on growing food for local consumption: Since no one has produced dicamba-resistant vegetable seeds, the farm had to close. Its crops could not survive the poison's onslaught. Dicamba is controversial even among row-crop farmers, since it's prone to lift off of one field and settle somewhere new—often the field next door, which may or may not be planted with dicamba-resistant seeds. In 2016 in Mississippi County, on a back road just south of the Little River Drainage District, two farmers fell into a dispute over the chemical. The fight ended in a series of gunshots. One farmer died, in what a judge later declared a murder. "People are just—they've just gone batshit crazy," a local farmer told me when I visited the region two years later. "I wouldn't dare go out and say there won't be more violence."

The one thing that does seem to survive this chemical are the weeds. By 2020, just five years after the dicamba-resistant seeds debuted, scientists found pigweed across several riverside counties that could endure the poison. Here was one more advance in the genetic arms race—a race that humans seem to be losing. "It's starting to get on that slippery slope of how much longer can we keep doing what we're doing and being successful," one of the researchers who found the dicamba-resistant weeds told an agricultural journal. For all that the country has gained, then, from the swamps we've drained, it's not clear to me how much longer the world we've built atop the former floodplain can persist in its current form. And it's not just dicamba. One of the more striking facts about the modern alluvial valley is that after we have spent years draining swamps, the ground here is now growing too dry.

Most of the rainfall comes in the off-season, so some farmers have installed pumps that suck water from deep underground. Arkansas now has more irrigated acres than Texas; in federal maps of irrigation, you can easily discern the outline of the alluvial valley. Many areas, especially those closest to the river, are wet enough that there's no reason to worry. But in parts of Arkansas, the water table has dropped a hundred feet. Since it's notoriously hard to measure the capacity of underground reservoirs, no one's quite sure how much deeper the water table can go. Already some of the local rivers and streams in Arkansas are running low more often, a signal that their water is seeping downward to refill the aquifer. This suggests that in the floodplain's remaining scraps of wetlands, the subsurface hydrology is changing, too, and the soils are growing drier—one more assault on what little is left of the swamps.

Researchers are scrambling for solutions, though more for the farmers than for the wetlands. Computer software can increase the efficiency of irrigation pipes. In places, the upper layers of the aquifer's sands have been intentionally exposed, allowing winter rain to slip underground more quickly. The most promising solution may simply be returning more water to the dried-out floodplain. The aquifers in the Yazoo Basin, too, were once in decline, but almost two thousand acres of new storage ponds have been built since 2010, which seem to have turned the problem around. On the western edge of the alluvial valley, near Little Rock, there is a massive pumping station—a three-story building, rising above endless fields, at the foot of a stunted little canal. The place has the feel of a bunker; the pumps are contained in concrete cells, below ground, where water drips from the walls. The idea is to take water out of the Arkansas River and send it through one more set of canals, supplying water to the nearby farms. After we've designed one system of artificial waterways to drain the region, in other words, we've designed another to turn it wet again. Work on the pumping station was completed in 2014, but due to a lack of funding, less than two miles of the canals has been constructed. So the station stands eerily quiet, empty and unused—a hedge against a precarious future.

A few times during our day together, as we crisscrossed the flatness of the Little River Drainage District, Boatwright repeated wisdom he'd learned from his grandfather. "Play the cards you're dealt," he'd said. *You can't undo the past,* in other words, *so take stock of what you've got, then do the best you can.* Despite all that's gone wrong in the floodplain—the suffering towns, the supercharged weeds, the dwindling aquifers—our hand is not all bad here. This land could feed the world, if only it were better used.

Still, for me at least, knowing the river's past made it hard to miss its lingering echoes. Driving along Highway 61—the famous blues highway—I could see old shacks and gas stations crumbling into the surrounding fields, a testament to the former boom years, and to the fact that the boom was over. (The old mansions built by the planters were clean and trim, meanwhile; you could rent one out to host your wedding.) Even the landscape itself seemed haunted by its former form: weeds sprouted from the puddles that spread over the turnrows—miniature swamps returned.

Toward the end of our tour, over a lunch of fried catfish, Boatwright noted that the system he runs requires constant upkeep. As mud and dirt flow off the farms in Missouri's bootheel, the ditches lose their carefully engineered grade. Every seven years or so, excavators must scoop out the collected sediment to smooth the bottom back out. If for some reason they stopped their maintenance, the mud would accumulate, until the ditches were plugged completely. Within ten years, Boatwright said, water would spread back over the land—its first step in becoming a mighty swamp once more.

It's a reminder that this land is not really land but a severed piece of river: a low expanse that was built by its flow, by its floods. This is the bottom of the continent, where the water pools, whether we like it or not. Ultimately, the fate of the floodplain—whether it stays farmed or goes feral—will depend on the river. And lately the Mississippi seems to be bringing more and more water.

The Low-Sill Structure at Old River (Concordia Parish, Louisiana).

11

The Great Unraveling

How engineering may fail in the challenging years to come

The wettest twelve months in U.S. history began in the summer of 2018, and by October, enough of the country's rain had drained south to Louisiana to put the Mississippi at flood stage. That winter and into the spring, the storms continued. Officials from the Corps of Engineers gathered every morning in a New Orleans conference room to review their flood-fighting operations—to discuss the latest forecasts, and to review notes from inspectors, who assessed every foot of every levee every day. The room was windowless, so the team's opponent was invisible. Still, it was hard to forget that just feet from the office, the river was snarling past, carrying millions of gallons of water each second. The central question for this team was which components of the Mississippi River and Tributaries Project needed to be engaged: which gates pulled open, which pumps turned on. The Bonnet Carré Spillway became a particular focus of the talks.

Its history offers a rough benchmark of the river's changing behavior. Over its first eight decades, the spillway—a key tool to protect New Orleans—was used sparingly, just ten times in total. Then the Mississippi seemed to shake something loose: the spillway had to be opened in 2011, in 2016, in 2018. Each of these openings comes at real cost. The spillway flushes its fresh water into Lake Pontchartrain, which disrupts the typically brackish ecosystem. The resulting oyster die-off can reach as far east as Alabama. Economists calculated that the opening in 2011, for the biggest of these floods, cost the Mississippi economy $58 million.

In February 2019, four months into the latest flood fight, the river rose and hit the trigger point again. The team of Corps officials had no option. For the first time, the agency engaged Bonnet Carré in back-to-back years. After several weeks, the flood crest passed, and the water subsided enough for the Corps to shutter some of the gates at Bonnet Carré. On April 11, the Corps closed the spillway completely, though officials kept a wary eye on the still-swollen river.

Just a month later, on May 9, they noted another crest pressing downstream. The leadership of the Corps's New Orleans District looked at the data and announced that in five days they would have to reopen the gates. "It's a bit extraordinary," the Corps's emergency operations manager told the press. Not just back-to-back years now: the spillway would be used twice in a single year.

Then came the rain, so much in one night that the river rose six inches. The discharge rate remained below the official trigger for the spillway's operation, but lately the river's behavior had changed: the same flow rates yield a higher stage. The water was predicted to reach 17.5 feet, higher than New Orleans had seen in more than forty years. For Corps officials, such a rise would bring the river too close to the tops of the levees, which sit at around twenty feet. At the next morning's meeting, they made the executive decision to open the spillway—immediately. Within hours, a crane plucked wooden pins out of the Bonnet Carré's gates. The water came roaring out, the collected runoff from a flooded continent.

The decision was a good one—for New Orleans, at least, if not for the oysters. The MR&T Project did its job that spring; all along the Lower Mississippi, the levees and spillways contained, once more, the latest great flood. According to Corps of Engineers officials, after eighty years, no levee built to the project's standards has ever failed. But the scramble at the spillway brings to mind an old adage, familiar to any river engineer: there are two kinds of levees, those that have failed and those that *will* fail.

That no system of engineering can last along this river should be clear to anyone who knows the history of the muddy terrain. Perhaps we should forgive the Corps of Engineers, though: for the entire first century that it spent renovating the river—even as Harley Ferguson began to snip apart the river's bends—no one had managed to untangle the mysteries of the river's past. Only in 1941, as the Mississippi River Commission was finalizing its plans for the MR&T Project, did the army assign a geologist the task of piecing together the alluvial chronicles. It's worth turning to the resulting study, because it touched on so many of the problems the river faces today.

Harold Fisk, the geologist the Corps hired, was raised on a fruit ranch in southern Oregon and had arrived in Louisiana a few years earlier to assist the state's geological survey. To great acclaim, he charted the ancient meanderings of the river across four parishes. Now in 1941, at the Corps's behest, he expanded his work to span the whole of the lower river. He sometimes chafed at his new employers, especially their militaristic discipline, which in his view bordered on unscientific: he noticed that as they studied new projects, they took borings at precisely repeated intervals, even if this meant skipping over essential details of the surface geology. At one point, Fisk traveled to Nashville and Frankfort, Kentucky, to gather data from the states' highway departments and geological surveys—which caused a furor, as these cities lay beyond the boundaries of the engineering office in which he was employed. When he was reprimanded, Fisk "reacted violently and with great profanity," a friend later remembered. Fisk made it clear that he would not abide by arbitrary bureaucratic rules. What he respected were geological boundaries. Despite his fury, in photos the geologist looks a bit more jovial than most of his army peers, and a bit less orderly: loose and smiling, a nutty professor. Eventually, he learned how to work with the army. Over the

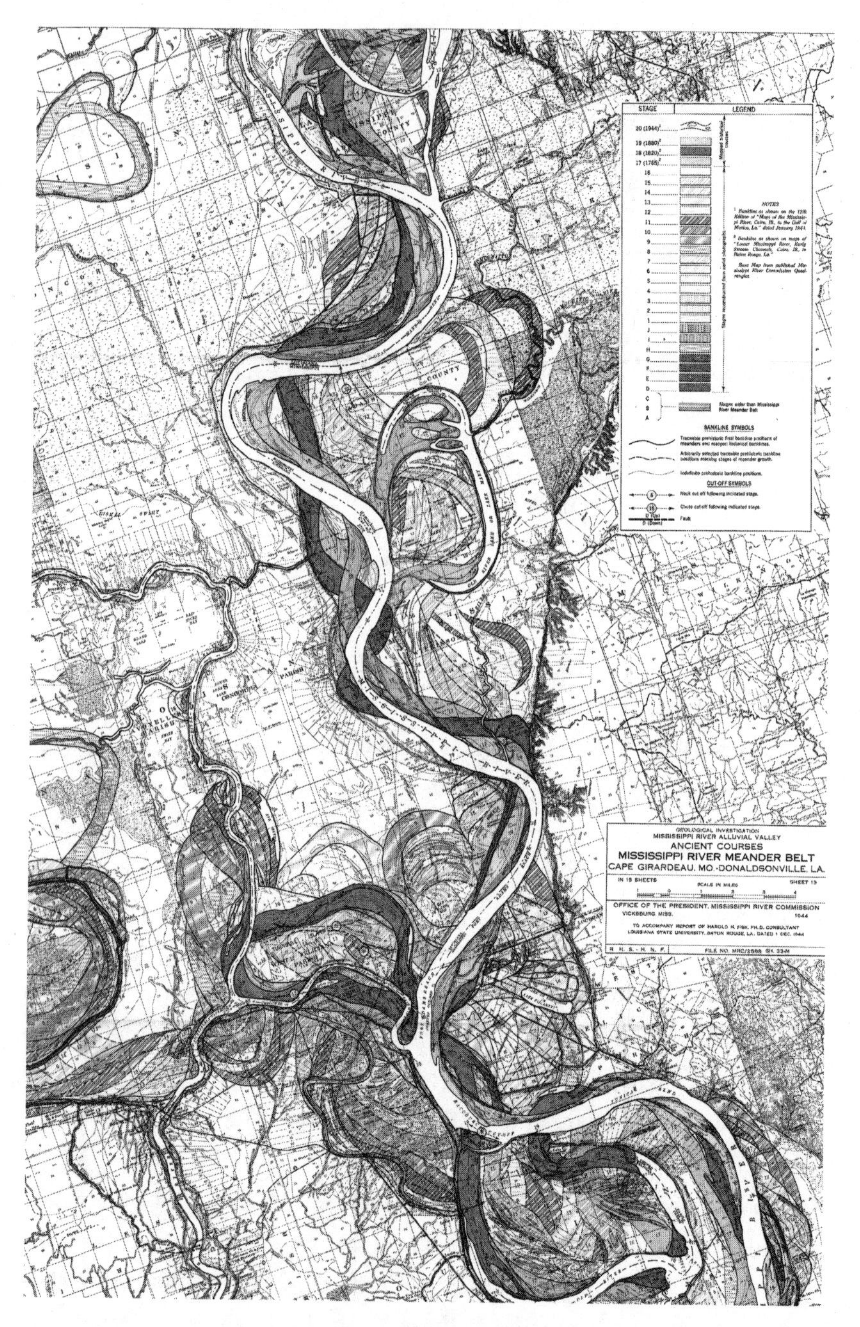

Fisk's maps, by documenting the river's former paths, helped reveal the extent of its meander belts.

next seven years, his team completed nearly fifty studies of Mississippi River geology.

Fisk knew that the river was shifty, growing its bends, carving its shortcuts. Still, most of the time these curls wind along a consistent axis, which means that at the outside of each river bend, where the laggard water cannot carry its mud, the soil accumulates into natural levees. These remain in place even as the bends shift and the river moves elsewhere; visible from above, they mark the boundaries of the "meander belt," the zone within which the river has curled. So Fisk looked first at the aerial photographs that had just begun to be collected. Then he and his team went out in the field to check that the vegetation and soil types matched what he'd presumed from the photos. Finally, he checked again by examining some sixteen thousand soil borings—which showed him the state of the mud as deep as thirteen thousand feet below the ground. Since the speed of a river helps determine what kinds of dirt it carries, the layers of sediment beneath the surface helped reveal how the river had shifted over time.

The report Fisk wrote became a classic—the first work to translate the local geology into insights useful to engineers. It's perhaps most famous for its maps, which are considered a landmark in cartography. In psychedelic ribbons of pastel colors, rippling and layered, Fisk showed the river's wandering years. The maps managed something rare: on a single, static document, the geologist emphasized that this place had always changed. He highlighted rather than obscured the wildness of the Mississippi.

Fisk's report concluded that throughout all these changes, the Mississippi River had remained a "poised stream." For thousands of years, ever since sea levels settled into place after the Ice Ages, its bed had neither risen nor fallen; its length had shifted, of course, but on average it remained roughly the same. Throughout various eras, the river built its natural levees to surprisingly consistent heights. This equilibrium yielded what Fisk considered the best-case scenario for flood control and navigation: a deepwater channel running from the top of the alluvial valley to the Gulf of Mexico. The river's poised status depended on its ability to meander, which helped it sustain a steady grade along the riverbed. But as Fisk wrote his report, the age of meandering was ending. The river had been chopped up, shortened, locked in place. He made passing references to changes he described as "artificial" but offered no prognosis of what these interventions might mean in the years to come.

Andrew Humphreys and James Buchanan Eads spoke with triumphal confidence of their ability to know what a river might do. Many modern hydrologists have adopted a much warier outlook: they consider river engineering to be a massive trial-and-error experiment, conducted on a geological scale. Once installed, infrastructure alters the flow of the river, which alters the flow of mud, which creates an entirely different river than the one just engineered. "Until an engineering structure is actually tested with real hydrologic events its effectiveness is only theoretical," Paul Hudson writes in *Flooding and Management of Large Fluvial Lowlands*. He offers the MR&T Project, and Harley Ferguson's cutoffs in particular, as, literally, a textbook example of the problem.

When Ferguson launched his program, the river was still adjusting to the surge of sediment that had been dumped into the river by the earthquakes of 1811 and 1812. Now he created another massive geological change: the slow and steady slope of the riverbed became punctuated by a series of sudden drop-offs, a sequence of sixteen underwater cliffs—knickpoints, hydrologists call them—between Memphis and the head of the Atchafalaya River. After the cutoffs were completed, the river sought its new equilibrium, through what few means were left to it. The Corps had to scramble to thwart its response.

First, the force of the water ripped Ferguson's small snips into deep and wide channels. This was the intended goal, but the excavated mud had to go somewhere, and the Corps found it was mostly reassembling into bars just downstream of the cuts. So the dredgeboats got back to work, now at a vastly expanded scale. In the 1950s the Corps resumed construction of wing dams, too, meant to "train" the river, using its own hydrological force to scour a deep channel. (This was a strategy the Mississippi River Commission had tried in the 1880s before giving up the next decade.) Soon it became clear that the deepened channel was undermining the integrity of the riverbanks: huge chunks of soil kept heaving into the river. For decades, the standard defense against erosion had been the mattresses made of interwoven willow boughs that the Corps anchored onto the banks with mud and rock. The engineers had long been searching for a better approach. Now the Waterways Experiment Station in Vicksburg began to double down on these revetment experiments.

Ultimately, the process was industrialized. Now each summer, when the river drops low enough to provide access to the banks, a matboat is sent downriver—a massive, mobile, floating factory, from which gantry cranes unspool modern mattresses. They're now made from concrete pads—each four feet by twenty-five feet—that are wired together with steel cables to form a 140-foot-wide flexible sheet. Draped along the riverbank, the mattresses conform to its curve and grade.

The original boat operated for more than seventy years, well into my era of river travels. When I arranged a visit, the crew was at work in Issaquena County, Mississippi—the least populated county on the east side of the river. So my arrival was the inverse of the typical river experience. Crossing the levee, I left behind the quiet countryside and entered a cacophonous zone of industry. On deck, men scurried atop an unspooling sheet of concrete, tying the wires with pneumatic tools that resembled jackhammers. Bulldozers were parked on the riverbank, providing an anchor. My ears, as required, were plugged against the noise.

This job is endless. After seven decades of work, with more than a thousand miles of riverbank lined in concrete, the project is incomplete. And the concrete must be maintained. Current rates of funding allow the Corps to repair less than one percent of the river's revetment each year, and still the matboat can be at work—ten-hour shifts, twelve days on, two days off—for months in some years.

The most worrisome consequence of Ferguson's cutoffs lies at the bottom of the river, where the river continues to respond to the knickpoints the only way it can: by slowly wearing them down to flatness. The top of the knickpoints is carved away, in an effort to match the depth downstream, an erosive pattern that gradually creeps upstream. Near Hickman, Kentucky, the river has scoured so much mud from its bottom that it's reached a dense layer of ancient clay—a hardpoint, as Harold Fisk dubbed the thing when he identified it in 1947. It has served as something of a wall, stopping the degradation from migrating farther north. During droughts, towboats run aground on this clay, so the Corps of Engineers has contemplated slicing through the layer to deepen the river. The agency's geologists have encouraged caution: remove the clay, and you could launch sudden and substantial changes for dozens of miles upstream, causing the river to carve a deeper bed as far as the Olmsted Lock on the Ohio River, which was completed in 2018 after thirty years of work and $3 billion in expenditures.

All the mud that the river scrapes from the top of the knickpoints is carried downstream, where it piles up as the river tries to balance out the difference. This increases the height of local floods. The impacts were clear by the 1970s, when the Corps of Engineers researcher Brien Winkley published a withering assessment of the cutoff program. "The river is not behaving as hoped," he wrote. Indeed, given the site of the worst aggradation—just downstream of where the Mississippi's great drainage basin squeezes into a narrow pinch-point—Winkley realized that a potential crisis loomed.

~~~

Just south of Natchez, the water running off the million-plus square miles of the Mississippi watershed passes through a bottleneck of lowlands that spans less than fifty miles across. This is seminal point for the modern river. Roughly six thousand years ago, the Mississippi "avulsed" here: it crashed through its self-built levee to forge a new distributary, which eventually became its main outlet. The river abandoned an old southbound route and veered east, toward Baton Rouge.

Over the next few thousand years, three more major avulsions
~~~

occurred, farther down the new route. The river switched through these forking paths, often splitting its flow. Each new distributary, meanwhile, branched into a series of veinlike channels. The mud that was carried down these veins blossomed into the hundred-mile-wide expanse that is the modern delta.* Call it the deltaic chronicle: you can read its chapters by tracing the old abandoned waterways that still reach toward the gulf. Bayou Teche and Bayou Lafourche and Bayou Plaquemine, all ancient channels of the Mississippi, have in the years since their peak dwindled into pliant little streams.

The pinchpoint near Natchez is also where the Mississippi's southernmost tributary, the Red River, dumps its water into the flow. Through the millennia, as these two rivers have danced their meandering dance, their precise point of intersection shifted. A few thousand years back, some of the Red River floods began to trickle into a little bayou, a meager distributary that ran through an old Mississippi channel. When the two rivers settled into their current formation five hundred years ago, that little bayou was relinked to the big Mississippi. It grew. The Atchafalaya, we call it today, a name that's come to dominate many apocalyptic visions of the river's future.

It's ironic that the Atchafalaya was formed as little as decades before Hernando de Soto stumbled into the watershed: the conquistador might have recorded grand changes on the Mississippi, if only he'd known what to watch for. Even centuries later, the Atchafalaya offered "a virtual laboratory for the study of deltaic processes," in the words of one geologist. This distributary emptied into an expanse that, thanks to a fluke of the arrangement of the river-built ridges, was all but completely cut off from the Gulf of Mexico, and that for millennia had been untouched by the Mississippi's sediments. So upon the birth of the Atchafalaya, its receiving basin was mostly occupied by one big freshwater lake—which now began to fill with mud. A miniature

* The delta is wider still if you include the Chenier Plain, which was built by gulf currents dragging the river mud west, forming a marshy coastline that stretches all the way to Texas.

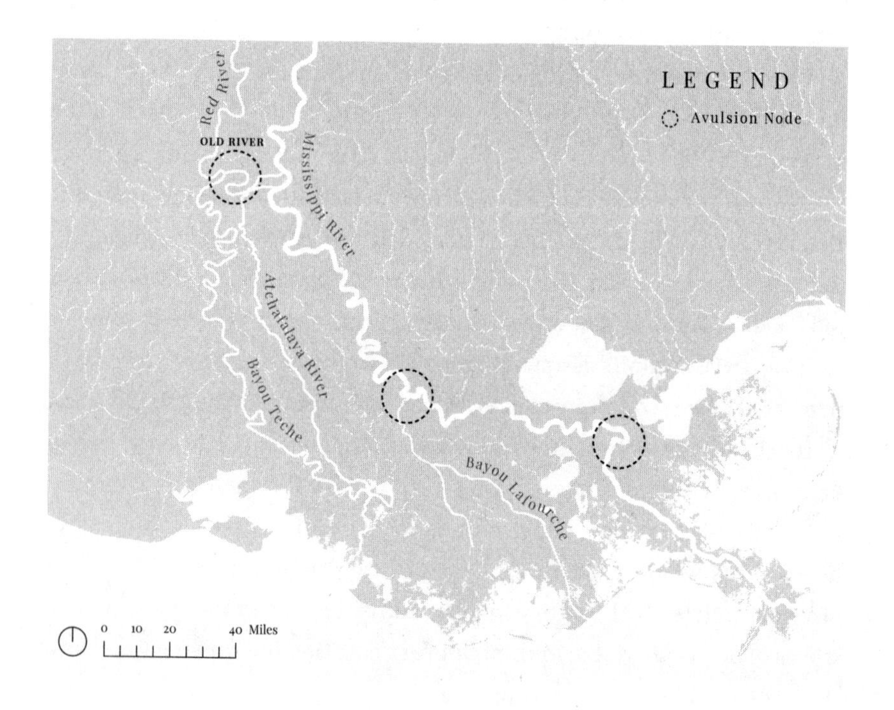

Sites of major avulsions in the delta.

delta appeared, tucked within a cleft of the larger delta. As colonists arrived, as old cultures shattered, a fresh landscape was emerging in the Atchafalaya Basin.

At the dawn of the American era, three centuries after its birth, the Atchafalaya was still not much of a river. After a few dozen miles, its channel frayed into a set of interlinked ponds and bayous that passed through muddy, tree-clad forests. The result was a maze where, as one visitor declared, "the inexperienced traveler would require the thread of Ariadne in order not to wander forever." Few travelers bothered, since forty miles of the Upper Atchafalaya were plugged with driftwood. When sugar boomed, though, rich men swooped in to set up plantations on the still-growing slivers of land within the Atchafalaya swamps.

The reengineering of this distributary began in 1831, when Henry Shreve carved his ill-fated Mississippi shortcut: the bend he lopped off included both the mouth of the Red and the head of the Atchafalaya. He did not seem to care that by putting the Atchafalaya's headwaters on an oxbow, his work would slow the distributary to a trickle. A few years later the Corps of Engineers concluded that removing the logjam would solve the problem—while also turning a half-million flood-prone acres in the Atchafalaya Basin into farmable land. But this was the era when Congress was loath to fund improvements, so the task was left to local engineers. The fire they set killed thousands of alligators, apparently, but failed to clear the underwater wood. Finally, a few years before the Civil War, a state-owned snagboat got the trees out. The Atchafalaya was set loose. The little river became a torrent.

Planters had been eager to remove the raft so that steamboats could carry their crop. But as the years wore on, they found they had no crop to carry. Their fields were mired by flooding. In the 1880s the Mississippi River Commission sent an army major to survey the region. He found that just the first thirty miles of the river were still occupied; beyond that, the plantations lay empty. There were, instead, plenty of "swampers" downstream—families, often French-speaking, who lived in floating homes or cabins erected atop the narrow banks of the bayous, and who made their money felling timber and catching fish. These swampers are what makes the Atchafalaya famous today: the bayous here are remembered as a beating heart of Cajun country.

The 1927 flood was the beginning of the end for even this watery way of life. The timber industry was already waning—the trees had been cleared, mostly—when the flood smothered ridges that in the swampers' memory had never gone underwater. And what came after the flood was even worse. The Mississippi River Commission had shuttered every other distributary but left the Atchafalaya alone, in part because studies suggested closing this outlet might lift river stages as much as four feet on the Mississippi. So the Atchafalaya was the one remaining alternate route to the ocean. The fact that the Atchafalaya's swampy forests were not farmed was even better: here, Edgar Jadwin decided, was the perfect site for the largest spillways. The distributary became the anchor of the Mississippi River and Tributaries Project. In the case of the project design flood, more than half the water—1.5 million cubic feet each second—is to be directed through the Atchafalaya swamps. That's more than twice the average discharge of the Mississippi River, a lot of flow for a distributary.

The Corps of Engineers had to renovate Atchafalaya once more. New levees arose along the river's first fifty miles, protecting the viable farmland. Harley Ferguson arrived with his dredgeboats to slice a second outlet through the ridge that separates the freshwater swamps from the Gulf of Mexico, ensuring that the torrent of diverted floodwater would not get trapped. Then he began to carve an actual river through the morass of swamps. The Corps erected a set of "guide levees" that run north to south, fifteen miles apart, delineating the path of the new floodway.

As these walls came up, the swamper families began to abandon their watery life, settling on the dry side of the levees—still living out of houseboats, sometimes, now installed atop pilings on solid ground. Today the dry side of the spillway levees is quiet, rural Louisiana in its standard form: sugarcane fields, humble homes, and trailers. The levee stands so tall—up to twenty feet above the natural elevation—that from the highway you cannot see into the swamps. Climb to the top, though, and you find yourself standing on the edge of a water world.

I know few places where the transition from one ecozone to another is so abrupt or complete. Beyond the guide levee lies the largest remaining freshwater river swamp in the country, nearly a million acres of the sort of cypress forests that once ran along the Mississippi as far north as Missouri. For fifty miles, there are no roads, just lakes and streams and forests that blare with birds. Crawfishermen still work out there; you can hire an airboat to cruise through the bayous and spy on alligators, or rent a kayak and wind through the trees. There are scattered camps on the ridges and riverbanks, and floating houseboats still. But an amateur anthropologist who collected the oral histories of the swampers told me the last full-timer who refused to leave the spillway passed away in 1995.

The swampers were leaving, often, because the Atchafalaya was becoming too mighty: the water, squeezed between the guide levees, rose. (It did not help that when it drained, the water left behind its mud in great smears.) As early as 1812, engineers had discussed the possibility that the Atchafalaya could steal the flow of the Mississippi River and become the site of the next great avulsion. In the 1880s, worried over this possibility, the Mississippi River Commission installed underwater dams to prevent the distributary's growth. When Congress debated how to respond to the great flood in 1927, the head of the commission suggested that the river was "just itching" to turn down the Atchafalaya. Nonetheless, the Mississippi needed a floodway. So the engineers went to work to ensure that the little Atchafalaya could accommodate the waters of half of a continent. This opened a bigger and more tempting path.

Finally, in 1945, with the new floodways still under construction, the president of the Mississippi River Commission thought to re-raise the old question: might the Atchafalaya River become the site of the latest avulsion?

Harold Fisk had not directly studied this issue, but he'd learned enough to have an answer. "There is," he replied by official memo, "a definite possibility."

For New Orleans, avulsion would present an upside-down catastrophe: the once-flood-prone city would sit alongside a shriveled little

Mississippi, a brackish creek that could not supply the fresh water the city needed for its power, for industrial facilities, and for its citizens to drink. The MR&T system would become a set of walls and spillways clasping a ghost of a river—the largest engineering project in U.S. history, still under construction then, would be rendered superfluous by a single disaster. So the Corps of Engineers took seriously Fisk's warning and asked the geologist to conduct a fuller study. Published in 1952, Fisk's second report concluded that river capture was more than possible: it was imminent, destined to occur as soon as 1965.

Today the two rivers' meeting place has been plugged with a dirt wall. Just upstream, 566 feet of the Mississippi River levee have been removed, replaced with eleven steel gates. Envision the world's largest garage doors: such is the new and human-made head of the Atchafalaya. The structure that holds these gates in place, pressed seamlessly into the levee, is wide enough to accommodate a two-lane highway across the top. When you drive along this road, a hundred-ton gantry crane looms over your car. Below you, alongside the intake channel, concrete wing walls flare outward, funneling the water toward the gates. Steel beams, driven ninety feet below ground—past all the river's mud, down to the underlying sand—help rivet the engineering in place. Just upstream a second line of gates, six times longer, looms over a patch of batture, ready to provide a second outlet when the river rises out of its banks. Construction on the complex began in 1955 and was completed by 1963, just two years before Fisk's projected avulsion date.

The Old River Control Structure, as this place is known, sits two hundred miles upstream of New Orleans, seventy-five miles above Baton Rouge, more than ten miles from any officially incorporated community, amid soybean fields and hardwood forests. There is no effort at beauty here, just the bare-bones brutalism of the engineer. "What we've done here at Old River is stop time," an army major general once told a journalist: the gates are intended to ensure the distribution of flow remains forever in the approximate ratio of the 1950s, when 70 percent of the water stayed in the Mississippi and 30 percent entered the Atchafalaya.

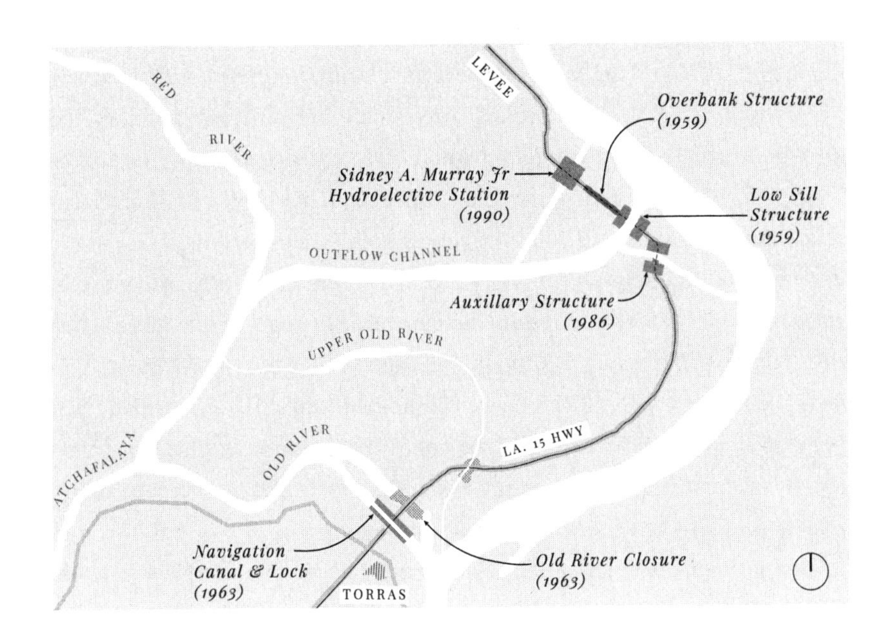

The engineering at Old River.

The fact that time can't ever be stopped completely became apparent to the south, in the distributary's swamps, where the mud-built delta continued to grow—now even faster, in fact, since the sediment was trapped within the artificially narrowed spillway. Two lakes in the Atchafalaya Basin shrank by sixty square miles in the three decades after the 1927 flood. By the 1970s, the amount of open water was a "tiny fraction" of what had existed before the flood, as one geologist noted in a report.

I visited Old River in 2019, to report on how the site fared through the long flood that spring. From a concrete catwalk built along the confluence, the Mississippi River looked like a varicose vein: the nation's central artery had grown so swollen that it was ready to pop. To the east, across the river, water covered five miles of floodplain, so far that I could not see the distant shoreline. Behind me, to the west, lay the concrete-lined channel that marked the beginning of the Atchafalaya. A siren wailed—a warning that one of six steel gates beneath my feet would soon inch upward, letting a trickle more water through. The water slammed against the gate, coiled into a whirlpool, then dropped nearly twenty feet into an engineered waterfall—connecting two rivers that, under natural conditions, should be the same height. The differential was nearly as large as the Corps of Engineers allows here; any more, and the force of the water could rip out the gates.

Ricky Boyett, the chief of public affairs for the Corps's New Orleans District, was my chaperone. On a press tour a few weeks earlier, deep in the bowels of a New Orleans pumping station, another reporter had asked him what the agency might do if these gates were ripped loose, and the Atchafalaya became the big river. Would the Corps try to reengineer the entire Mississippi back to its familiar shape? Boyett was glib: "You know, Army Corps—we can," he said. Then, laughing, he backtracked. "I always say we've picked the worst enemy in the world, and that's Mother Nature," he said. "Because at the end of the day, we know that Mother Nature can do what Mother Nature wants, and we're just trying to hold it on pause."

~

The pause at Old River was nearly a short one. As soon as the gates opened, the ceaseless rush of water chewed into the outflow channel; by 1964, the resulting scour holes were already forty feet deeper than the pilings. That same year eight unmoored barges were lured by the current and slammed against the structure. The complex had to be closed for repairs. The water stacked up against the shuttered gates for months, growing into a taller and taller wall. When the complex was opened again, the unleashed flow added to the existing scour.

More barges hit the gates the next year. The Corps assigned a picket boat to patrol the water, watching for rogue vessels. The structure emits a warning tone, too, so pilots cannot forget they're near a precarious and dangerous site. The first expedition I joined with John Ruskey launched nearby; we could hear the siren as soon as we reached the Mississippi, a single electric note, plaintive, that echoes every few seconds across the surrounding lowlands. When I flipped through Ruskey's official navigation maps, I noted the warning overlaid on the chart at the intake channel: "Under no circumstances should any vessel attempt to enter."

Water does not heed such instructions. In March 1973 the river surged into its latest flood: two million cubic feet of mud and liquid pulsed down the Mississippi every second. Pressed against the gates at Old River, the water grew as tall as a six-story building. A quarter of the flow plunged downward through the control structure, through the gates, and as the water bounced and jostled, it set the steel and concrete aquiver. Anyone who dared to drive out onto the road atop the gates and park their car would find that as soon as they opened the door, the shaking would knock it shut again. Over the next few weeks, the rush of water grew and grew.

Near sundown on a Saturday in April, the facility's foreman stepped outside to survey the conditions and noticed that one of the wing walls was missing, presumably ripped free and swallowed by the flow. He called a disbelieving colleague on the office radio. "Kid, it's gone," he said.

The Corps of Engineers shifted into emergency mode. Louis Logue, who worked as the head of a design unit at the Corps, told me he had crews dumping boulders into the water, hoping to plug the hole that had carried away the wing wall. He could see that again and again, as the boulders tumbled into the water, they popped up on the far side of the gates, in the new mouth of the Atchafalaya. The Corps scrambled to get its massive river model outside Jackson, just retired, back online. The two remaining staff members kept it running twenty-four hours a day for the remainder of the flood, assessing what levees were at risk. But one key question—what would happen if the Morganza Spillway were opened for the first time—could not wait for laboratory results. Within days, the gates creaked open. As the pressure lessened at Old River, the Corps scrambled to initiate repairs.

After the flood ended, engineers drilled a hole beneath the Old River gates and dropped a camera into the murk. They found fish swimming where the pilings once stood. This fortification, they realized, was critically wounded. Designed to handle a thirty-five-foot head differential, now it could handle just twenty-two feet. The engineers had to install one more set of gates at the complex. The "auxiliary structure," as these gates are collectively known, quadrupled the overall price of the project, but did not solve the full problem. The long meandering dance of the rivers has laid down a layer of sand here, rather than the thick muck of backswamp that exists in places where the channel never touched—offering an easy site for the river to break through in an avulsion.

As the auxiliary structure was still under construction, Brien Winkley released his analysis of Ferguson's cutoffs. He identified one more problem: much of the mud being scoured off the knickpoints was piling up in the Mississippi, just downstream of this confluence. The Atchafalaya, meanwhile, was still carving a deeper channel, creating a head differential between the two rivers that could test the limits of the damaged structure. "The problems in the river below Old River could create a disaster," Winkley wrote.

What could be done? He suggested the Corps should study an even bolder program: shutter the Atchafalaya completely, forbidding its future theft.

That has not happened, perhaps because by the time the auxiliary structure was completed, in 1986, the river had settled into a quiet era. The Bonnet Carré Spillway had to be opened just once per decade—in 1983, in 1997, in 2008—which was a relief for the battered gates at Old River.

Then came a big flood in 2011, and the decade of big water. It wasn't just the size of the floods that was unusual; the timing of the high water became harder to anticipate. "We don't have normal river years anymore," as one Corps of Engineers official has put it.

A few years into this new era, a group of scientists decided they ought to consider the wider context, beyond the several centuries of notes jotted down by settlers. So they came up with a novel study: they examined soil samples collected from lakes alongside the Mississippi River and tree cores pulled out of oaks in the floodplain in Missouri. With this data, the group assembled a five-hundred-year-long record of flood events. The results showed that the recent rash of floods was less an anomaly than an intensification of an ongoing trend: flood magnitude and frequency had clearly and consistently increased for a century. The authors of the report did not hesitate to name one key culprit: the engineering—the levees along the river, the dikes in the channel. Think of these structures as plaque lining an artery. They plug up the flow, though in this case instead of a heart attack, the symptom is a flood.

Changes beyond the levee system, outside the wall, have added to the problem. Natural ecosystems act like sponges, soaking up water, then slowly releasing it into rivers and aquifers. But pavement is nonabsorbent, so every new parking lot, every new building, means more water rushing into rivers. Many farmers have installed underground "tiles" that allow their fields to dry quickly by shunting the water into their drainage ditches. But as a result, even where soil remains in place, the rain drains more quickly than it used to. Within the last century, the Mississippi's flow has increased by nearly a quarter in Minnesota; the summertime discharge in the Wabash River, in Indiana, has increased by a third; the Minnesota River's flow has *doubled.* All these rivers point toward the

Lower Mississippi, where since the 1950s, the discharge has grown an average of 4.5 percent each decade. It's slowly becoming a river that our levees cannot contain.

Climate change adds new wrinkles to these problems. The most famous consequence of a warmer atmosphere is the rise of the oceans—which certainly poses a problem for the river's delta, in Louisiana. But the new climate may cause struggles upstream, too. Warmer oceans evaporate more quickly, which means there's more moisture in the air. When that wet and heavy air hits land, it can be released in a heavier blast of rain. That sends more runoff into rivers, which leads to bigger floods.

The 2019 flood—the flood that forced the Corps to twice open the Bonnet Carré Spillway—offers a potential preview. That March a bomb cyclone walloped the Great Plains, delivering a blast of warm rain that melted a thick layer of snowpack. Since the ground below remained frozen, the meltwater had nowhere to go but into the local rivers. Soon every waterway in eastern Iowa, northwestern Illinois, and northeastern Missouri hit flood stage. Records broke at more than forty gauges. Near Lynch, Nebraska, a ninety-year-old hydroelectric dam crumbled under the weight of the water, releasing an eleven-foot wave that carried ice floes as large as cars. This water smashed through a highway and the pipes that provide Lynch with drinking water. Roughly a third of the town's homes flooded; every road out of town was submerged. The water, as water does, kept rolling downstream.

The various tributaries that drain much of this region eventually merge into the Missouri River, which soon set stage records at sites downstream of Omaha. "We were making what I consider life-safety decisions on an hourly basis," John Remus, the Corps of Engineer's chief of Missouri River Basin water management, later told me. Remus is tasked with managing the six massive dams on the Upper Missouri, which function as the country's largest water-storage machine. The lowermost dam, Gavins Point, has the smallest reservoir. The sudden cascade of water filled it within hours. Remus had little choice: if he did not release the excess water as quickly as possible, he risked a blowout. The question was how quickly the water could be drained without damaging the infrastructure. "It was an intense twenty-four hours," he said.

After the water poured out of the Gavins Point Dam, it ripped through Hamburg, Iowa, a town whose eleven hundred residents had been begging for years for help updating their levee. Fourteen levee systems failed above Kansas City; by the end of March, nearly five million acres along the Missouri River lay underwater. Roads were covered. Homes and grain bins were destroyed. Dead cattle floated atop the dark pool.

Too much water was coursing down the Upper Mississippi, too. By June, governors in eleven states had requested federal-disaster funds in response to the flood. The requests covered four hundred counties, from Nebraska in the west and Minnesota to the north all the way down to the Gulf of Mexico, where the water coursing out of the Bonnet Carré Spillway sent salinity plunging nearly to zero. The brown shrimp population in the Mississippi Sound was down more than 80 percent compared to five-year averages. Commercial blue crab landings, too, dropped 25 percent. The discovery of hundreds of dead bottlenose dolphins prompted the National Oceanic and Atmospheric Administration to declare an "unusual mortality event." By late June, an algal bloom, fed by the fertilizer carried downriver, had spread across Mississippi's coastline. Because the algae have the potential to produce toxins, the state closed its twenty-one beaches. They stayed closed all summer, devastating the economy of a region that depends on tourism.

The nation's twelve rainiest months finally ended in June, but the floods on the Missouri persisted into fall, in part because the reservoirs in the Dakotas were so full. To open up new storage space, in anticipation of another wet winter season, John Remus had to send a huge dose of water downstream. One farmer in Missouri could not reach his flood-seized home until Halloween. Finally, in December, the Corps of Engineers' Kansas City office declared that its flood fight was complete.

The one region that fared well throughout the 2019 flood was, ironically, the big southern river, where the MR&T Project has dutifully protected the floodplain farmland for eight decades. The system has never been pushed to its limits. The Morganza Floodway, the last outlet to be completed, has been used just twice; the West Atchafalaya Floodway not at all. The Mississippi River Commission likes to tout the tremendous return on investment—on the order of $95 saved for every dollar invested, it says. But the expanse of flooding in 2019—and the length of the floods, which shattered records—should be read as a warning shot. The river is not behaving as it has before.

It's not just floods. Some computer models show that in the years to come, the river's discharge might actually decrease due to less snow on the plains, creating more years of drought. Indeed, as if to affirm the fickle nature of our new climate, the rains all but ceased in the years following the big flood. By 2022, some sites on the river were setting record lows. Places where the towboats were accustomed to passing through forty feet of water now featured just ten. "People's backyard pools are deeper [than the river]," one captain told *USA Today.* For months, the Corps of Engineers had three dredgeboats near St. Louis working nonstop to keep the nine-foot channel intact. Cargo loads had to be decreased to ensure that barges would not run aground on the shoals. As the navigable channel narrowed, fewer barges could be lashed to each tow. Cargo rates skyrocketed; some farmers reported paying nearly four times the standard price. Some estimates put the cost of the drought as high as $20 billion. The next summer, too, featured a startlingly low river—so low by late autumn that the flow was insufficient to counter the inland creep of saltwater near the river's mouth, where the bed lies below sea level. This "plume" reached water intakes in Plaquemines Parish, tainting drinking supplies. Such droughts, if they recur, could be a fiasco for the navigation industry: as hard as it is for engineers to stop floods, stopping droughts is harder. Short of installing new reservoirs, we can't simply conjure more water where it refuses to arrive.

Calculating the intertwined effects of climate and land use and infrastructure is complicated, but models can give some sense of the coming flood regime. Two Degrees Adapt, a consulting firm working on behalf of a coalition of mayors along the Mississippi River, adjusted the Corps of Engineers' streamflow models to match the climatic change expected over the next several years. The resulting figures suggested that within a decade, some levees within the MR&T system are all but guaranteed to fail in a hundred-year flood. In a separate study, a group of academics found that in a future with continued high levels of carbon emissions, the Corps of Engineers' "project design flood" would turn from a thousand-year event to a thirty-one-year flood.

The Corps of Engineers itself has not published any in-depth studies that examine the MR&T system's prospects in a warmer climate, but a team of the agency's hydrologists have tinkered with these models. "Results indicate that the hydrologic conditions of the Mississippi River are not stationary and that discharges associated with extreme events are projected to increase in the future," the scientists concluded in a short paper. "This is an important finding in terms of the design and operation of the MR&T System." A less anodyne way of putting it: the world we lived in when we built these levees is not the world we live in now.

Indeed, weaknesses in the MR&T Project are already becoming clear. The accumulation of mud downstream of Old River means the same amount of water yields a higher river; in order to avoid exceeding the maximum twenty-two-foot head differential at Old River, the Corps must constantly adjust its gates. As the mud stacks higher, the problem will grow worse. And models show that eventually the sediment will accumulate further downstream, near New Orleans—enough mud and sand to lift the annual high water by as much as ten feet. These models did not account for the use of the spillways, but such a dramatic rise suggests spillways alone may not be enough. Dredging might provide some relief. So might sending more water down the Atchafalaya, though only at the price of making the distributary an ever-more-tempting path.

Farther north, where there are no outlets for the floods, the only real option in the face of rising water may be once more remembering that rivers need room. The Mississippi River City and Towns Initiatives, the coalition of riverside mayors who hired Two Degrees Adapt, has already made "green infrastructure" a key policy plank. Marginal land in the floodplain was once viewed as a potential source of tax revenue if it could be protected and developed, but it is now understood as a key tool for holding excess water. Even wetlands built on the "dry" side of the levee can store rainfall, slowing its flow into rivers. They can soak up fertilizers running off farms and trap greenhouse gases. The Nature Conservancy, the world's largest environmental nonprofit, has set a goal of restoring 750,000 acres of forests and wetlands within the historic floodplains across the Mississippi Basin. One conservation biologist from the organization told me that targeting the least desirable farmland—the sort of flood-prone properties that were never particularly productive—might yield a million acres in the alluvial valley alone, with little effect on our agricultural economy.

An even more ambitious approach would be to pull back the levees, widening the rivers' domain. This has already happened in Atchison County, in the northwestern corner of Missouri, where a stretch of levee was "obliterated" during the 2019 flood, to borrow the word of one Corps of Engineers official. Afterward, rather than rebuild at the same site, the local levee board decided to negotiate with landowners so the levee could be rebuilt farther from the river, reducing the pressure against the wall. This is not only good for human beings: on a similar setback project nearby, Corps of Engineers fish surveys found a record-breaking number of juvenile sturgeon.

But these are not easy projects. The Corps of Engineers agreed to the Atchison County setback only because the levee was so damaged that it was cheaper to rebuild entirely than to repair. The local levee board would like to see more levee setbacks at other, nearby sites. But the Corps will not foot the bill to move a levee—which costs as much as $24 million per mile—that is nominally doing its job. And the amount of floodplain reconnected here amounted, in total, to a little more than a thousand acres, enough space to hold far less than one percent of the waters of the 2019 flood. Undoing the work of the past few centuries, pulling back the levees to accommodate the coming floods, could cost us hundreds of billions of dollars.

Delta bayous (Terrebonne Parish, Louisiana).

12

Beautiful Country

Can unleashing the river revive its old domain?

One summer morning, as we wove in his skiff through the marshes on the east side of Plaquemines Parish, councilman Richie Blink told me that the federal government had recently deleted thirty-odd names from local nautical maps. Fleur Pond, Dry Cypress Bayou, Tom Loor Pass, Skipjack Bay: these were now undifferentiated, unlabeled expanses of open ocean. Such is the famous trouble of the Mississippi's delta, which is among the fastest-disappearing places on Earth. Louisiana loses a football field worth of wetlands every hundred minutes; over the past nine decades, a land mass the size of Delaware has disappeared.

Now it's also the site of what some have called the country's largest-ever ecosystem restoration project. The idea is to unleash the river—to tear down the levee, at least in one small stretch, so the mud can rebuild. It's a transformative concept, a step back toward the river's wilder day. But it's not clear that this project will be enough. As the seas rise, the delta seems all but guaranteed to be swallowed.

Blink grew up in Plaquemines Parish, which encompasses the wispy land along the Mississippi's final seventy miles, downstream of New

Orleans. He graduated from high school a few months before Hurricane Katrina made landfall. After the storm, he found work running boats to offshore oil platforms and quickly rose from deckhand to captain. "In the meantime, I was doing these guerrilla restoration projects between my shifts," he told me, "planting thousands of trees at a time." It was a way to fight his sense of powerlessness—against the storm, against his disappearing homeland. Eventually it led to a career change: while working with a nonprofit promoting restoration, he began running ecotours. The side hustle grew into a full-time business. Eventually, to better fight for the delta, Blink decided to run for parish council, too.

Our destination was a site on the river's east bank, where in the 1920s officials from New Orleans tore down eleven miles of the levee, restoring the riverbanks to their natural height—an early effort to dampen floods by giving the river more room. Since then the Bohemia Spillway has granted floodwaters an exit forty-four miles above the river's mouth. In 2012, with the riverbanks already weakened due to the major flood the year before, the annual rise ripped open a crevasse. A new distributary emerged. The resulting channel of water is now known, in honor of its rough date of construction, as Mardi Gras Pass.

Blink steered his skiff, the *New Delta,* into the pass, zipping past the leaning willow trees, then swerved through an intersection, weaving through a brief maze of canals and bayous. We blazed past the trees and the reeds and into an open bay.

Suddenly, Blink stopped the boat. To my surprise, he hopped overboard. Rather than sink, he stood. The water lapped at his calves.

Here, hidden below the surface, new land was growing: a muddy shoal, deposited by the muddy water pushing through the new pass. This was what Blink had named his boat for, he said. At night, he takes the *New Delta* out and runs in loops, almost hoping to run aground. He grants names to the shoals he finds—Turtle Island, Manatee Island—to replace those that have disappeared.

This, he says, is what can happen if you set a river free.

Not everyone is so enthused.

A few weeks earlier, on another visit to Plaquemines Parish, I had found Acy Cooper, the president of the Louisiana Shrimp Association, repairing his boat in Venice, the southernmost harbor on the Mississippi River. "There's not a son of a bitch in this parish, or within this industry, that doesn't want coastal restoration," he said. But there are other forms of restoration besides tearing down levees. When I mentioned my plan to tour the delta with Blink, Cooper told me that the young councilman could "kiss my ass. For him to sit here and go against everybody that he [has] ever known or ever looked up to." Blink was the sole member of the council who had not voted to condemn the plan to unleash the Mississippi River.

To grasp the disagreement, it helps to consider the local geography. A strange fact of the big, wide batture—the walled-in wild, as I sometimes think of it—is that it is an artifact of the alluvial valley alone. Downstream of Baton Rouge, in the delta, the river stacked its sediment in relatively straight lines, meandering only rarely; the levees clasp the river tightly, typically standing less than a mile apart. The batture in the delta, then, is rarely more than a thin line of trees. Most settlement is perched along the river's natural levee, which by Plaquemines Parish has narrowed to a thin ridge. That can make for a grim and hardscrabble landscape. The wind carries coal dust off stacks at the refineries. Gas pipelines finger up from the ground. As you head south, exurban mansions give way to trailers that have been jacked up on pilings. Stacks of riprap fortify the roads to the docks and marinas. Venice, at the end of the road, consists mostly of row upon row of utilitarian cabins built to accommodate the sportsfishers who flock here, seeking easy access to the marsh and ocean. That's where what wildness persists down here lies: not between the levees but on the other side of the wall, beyond this thin strand of settlement, in the vast domain of swamps and marshes that spread between the river's old routes to the sea.

The floodwaters that poured over the river's natural levees didn't stack the mud high enough to break the ocean surface, but they did reach close enough to allow grasses to take root. When the grasses died, they decayed into more soil, slowly building a precarious landscape where sometimes mats of plants float loose atop the water. In the marsh, as everywhere in the delta, the accumulation of fresh sediment lays a heavy weight atop the older soils below; slowly but inexorably, the ground compresses. As the river continues to deliver new doses of mud, the lost altitude is replaced, though eventually the river avulses, turning its attention elsewhere. Then the old riverside ridges sink; the surrounding marshland fragments. Tides capture the heartier bits of sand in the soils and rework them into barrier islands. Somewhere else, meanwhile, the river builds again.

Or that's how it used to work.

The river's current mouth is known as the Birdfoot Delta, since seen from above it resembles a chicken's foot, with three branches clawing out into the sea. It was still growing when the first French settlers arrived, and the colonists could mark its progress by noting how far the thin ridges of land kept extending beyond a fort they'd built. In the nineteenth century, the growth was abetted by the settlers rushing over the Appalachian Mountains and hacking down the forests that anchored the soils along the Mississippi's many tributaries. The agricultural boom cranked up the river's load of mud to twice its precolonial concentration, according to some estimates. The webbing of marshland along the river's final miles blossomed massively. But the boom could not last.

In 1897 Elmer Corthell—the engineer who had first drained Illinois's floodplain, then came south to work with James Buchanan Eads on the jetties—published a short piece for *National Geographic* about the decline. He noted that in the two decades since the jetties project, an old Spanish magazine near the river's mouth had sunk a foot into the sea. Corthell presumed the cause was the levees, which kept the river from rolling over its banks and feeding the marshes with mud. Now, instead, the sediment stayed in the river, running down the passes in the Birdfoot Delta, then tumbling off the continental shelf into a deep abyss.

Not that Corthell was especially worried: having the mind of an engineer, he saw nature as a resource to be shaped into something more valuable. Money was the metric that mattered, and the levee was worth the cost, since it protected farms. Thanks to these farms, the region would eventually be able to pay for a vast new levee system, he believed, one that encircled the marsh entirely, warding off the encroaching sea even as the land sank ever deeper.

So it went for several decades: Writers occasionally noted the sinking landscape. But no one ever wrote with any sense of panic, not even when it became clear that New Orleans itself was descending along with the rest of the mud.

Harold Fisk noted that local engineers had sped along this process, thanks to the miles of canals in the city, which, like the drainage ditches to the north, were built to tame the swampy terrain. The canals often sat higher than the city itself, so pumps had to lift out any accumulated rainfall. This whole process meant water never soaked into the ground. So the water table fell, which hastened the slumping of the ground. By the 1970s, in suburbs east of the city, the sinking had grown so severe that gas pipelines ruptured, which led to explosions that consumed several homes. Today around half of New Orleans lies below the level of Lake Pontchartrain. The only thing that keeps the city from joining the lake is a levee along its edge—something like what Corthell suggested, though this wall is just for the city and its suburbs. The marsh, for the most part, was left to wither and die.

In the 1960s, Texas officials decided that they wanted to divert the Mississippi's water west, toward the state's drought-stricken plains. Since this would reduce the river's flow through the delta, a team of researchers at Louisiana State University, led by geologist Sherwood Gagliano, began to investigate potential side effects. But when Gagliano noticed how quickly land was disappearing, that became his primary concern. His first report, released in 1970, provided a much more accurate—and alarming—picture of the decades-old subsidence crisis. The delta, Gagliano realized, was disappearing at a rate of 16.5 square miles per year. And the loss was not just in the Birdfoot Delta but everywhere along the coast. The report noted that the marshland dampened the fury of incoming hurricanes, so such losses mattered not just for the ecology but for everyone who lived in southern Louisiana.

Like Corthell, Gagliano blamed the levees—but not the levees alone. Over the preceding century, developers had bought up marshland parcels, ringed them in embankments, then installed pumps to remove the water. Their plan was to launch farms and neighborhoods on these newly dried plots. When hurricanes struck and filled the plots with water, the developers abandoned the projects. Gagliano's report noted that the resulting quilt of rectangular lakes offered "mute testimony" to these failed schemes. Another problem was the network of canals—"innumerable," as the report put it—dredged by oil companies to reach their wellheads and clear pathways for their pipelines. "These invariably alter circulation patterns in the estuaries, resulting in a general saltwater encroachment of the brackish swamps and marshes," the report explained. A subsequent study suggested that for every acre dredged in the marsh, nearly three more would be destroyed due to the resulting influx of salt. Still, when it came to solutions, Gagliano's report focused on the levee: cut it open and release its muddy water, Gagliano suggested. Ever since, engineers and ecologists have been captivated by the idea of such a "sediment diversion."

They quickly got to work testing the concept, slicing small and exploratory cuts through the rivers' natural banks near the mouth. (The land here, along the final dozen miles, was so meager that levees were never built.) When Gagliano released his study, the Corps of Engineers already had plans to build a more substantial diversion at Caernarvon, a site just upstream of Plaquemines Parish. The project's official purpose was to supply fresh water, returning the salinity in the local marshes to historical levels. But when construction commenced in the late 1980s, local newspapers described the Caernarvon diversion as a potential conduit for sediment, too—a way not just to preserve marsh but also to rebuild it. Powerful landowners pushed for an increase in the amount of water released by the diversion, so it might deliver more mud. Just a few years after the water began flowing in 1991, hundreds of acres of new marsh had formed, suggesting that diversions were a viable solution for the coast.

The federal government had begun to fund other restoration projects, too. Soil was dredged out of the river and dumped on the coastline. Rock walls appeared along eroding beaches. New sand was added to the barrier islands. These projects wound up being akin to the early years of U.S. levee building: they were a patchwork effort at best, stitching together small pieces of a landscape that was constantly eroding. The response never matched the scale of the problem.

What woke the world was Hurricane Katrina, which ripped through the marshland and put much of New Orleans underwater. Months after the storm, the state of Louisiana launched a new agency, the Coastal Protection and Restoration Authority (CPRA), which combined the missions of flood protection and coastal restoration under one literal and figurative roof. Flood control would no longer mean building just "gray" infrastructure like levees and sea walls; it would encompass the "green" infrastructure of helpful ecosystems, too. The agency, then, sought a different outcome than the Corps of Engineers' longtime mission to close off the river: central to the CPRA's plans was the construction of sediment diversions—the unleashing of the flow.

One might see the formation of the CPRA as a breakthrough moment. Finally, wealthy property owners realized they depended on the river's natural ecosystems. What was good for the monied class was good for the river, too. Another way of looking at it, though, is as one more case when the needs of the wealthy trumped the needs of everyone else.

The year after the CPRA was formed, agency officials approached leaders in Plaquemines Parish to discuss a proposal for a large-scale diversion. The state wanted to situate the spout near the community of Myrtle Grove, a collection of luxury homes on stilts above the marsh in Barataria Bay, on the river's west bank. You might think these homeowners would be delighted to know their shriveling marshland—some of the worst hit in the region—could be saved. But the local reaction, from not just the homeowners but the fishing industry, the local oil companies, and the farmers who grow citrus atop the ridge, was a near-unanimous no thanks. Nearly two decades later, Acy Cooper retains that same stance.

The proposal to build new marshland is a rather radical change from the historical precedent. The first European settlers were often overwhelmed

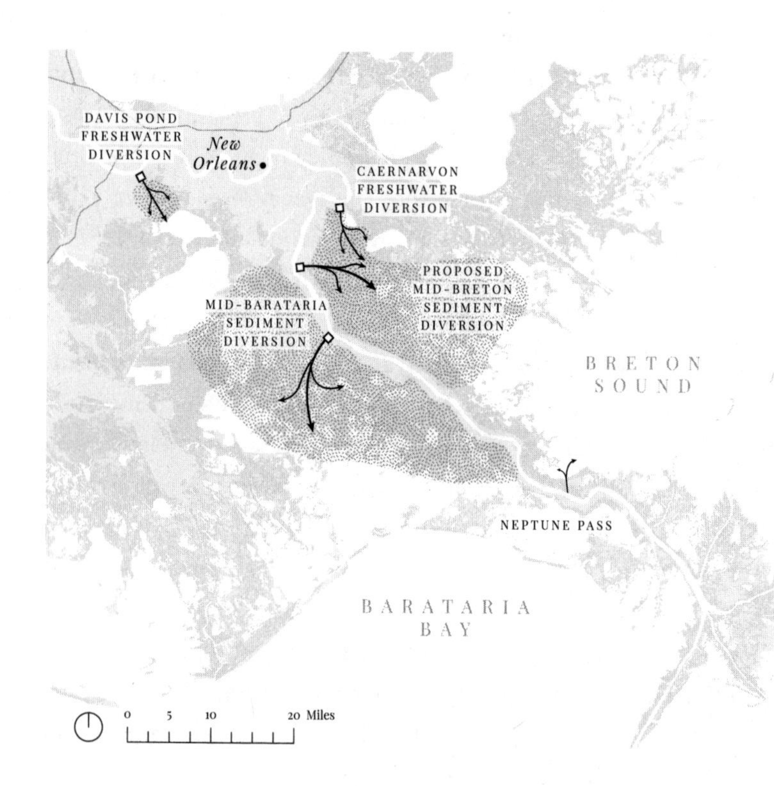

Current and planned divisions in the delta.

by the delta, which was far bigger and muddier than any wetland they'd known. So they tended to dismiss its worth. One eighteenth-century French cartographer depicted a large swath as a blank mass, labeling it as "trembling land and swamp." More than a century later, an army surveyor made his opinion even clearer by pointedly declining to enumerate the marshland's features. A list of the "multitudinous islands and sheets of water would add nothing" to his description of Plaquemines Parish, he wrote. The monied settlers mostly preferred the narrow ridges. So the marshes and swamps became a no-man's-land—or perhaps it's better to say an every-person's-land.

The people who inhabit these swamps offer a contrast to the planters and the engineers who dominate the story of the great river. At times, marshland people actively resisted these men's collective reigns. In the late eighteenth century, for example, Juan San Malo fled slavery on a plantation upriver of New Orleans and became a ferocious enemy of the Spanish government. Once, according to colonial records, he drove an axe into a tree at the edge of his territory. "Malheur au blanc qui passera ces bornes!" he proclaimed—*Woe to the white man who would pass this boundary!* His compatriots are said to have roared their approval. They never traveled without arms and did not shirk from a fight when colonial authorities attempted raids.

The marshland itself was essential to this campaign of resistance. San Malo's followers—fellow runaways, who numbered as many as a hundred—lived in cabins in two permanent settlements. One could be accessed only by wading through chest-deep water, pushing through reeds too thick for a canoe. Atop the stubby bits of land in the delta, the runaways grew corn and squash; on the water, they trapped birds and fish. They collaborated with people who remained enslaved, who would drive their masters' cattle to the edge of the swamps, where they could be easily stolen and slaughtered. San Malo wandered, living itinerantly, checking in on the members of his small society.

Eventually, San Malo was betrayed and captured, then hanged. According to the songs sung to remember his legend, his body was left exposed on the gallows tree to be slowly eaten by crows. Perhaps the white men did not know that in Bambara culture, a violent death and a long-hanging body give the victim stronger magic in the afterlife. And indeed, San Malo's execution failed to stop the runaways. Six months after the hanging, the very man who had stirred up public furor over San Malo found that eight of his slaves had disappeared. Many of the planters resigned themselves: it was not worth the money to pay for slave-hunting expeditions. The marsh was out there, undrained and undrainable. For years afterward, the river's southern forests were dotted with runaway camps.

Immigrants from the Canary Islands settled in the marshes near New Orleans in the same era. A few decades later the pirate Jean Lafitte used the coil of waterways to the south of the city as a smuggling route for contraband. Displaced Indigenous people began arriving in the delta, adding to the existing Indigenous villages. They figured it was better to build a home here, beyond the notice of settler culture, than to be forced west by Andrew Jackson's armies. Filipino immigrants took up residence in stilt villages raised ten feet above the water, where they processed the shrimp they caught by dancing atop the shells.

The most famous delta people arrived in the late eighteenth century: peasants of French extraction, known as Acadians, whose ancestors had been evicted from Nova Scotia at the outset of the Seven Years' War. Eventually the Spanish crown invited them south, to fill up the fledging colony of Louisiana. Through the lazy habits of the local tongue, this branch of Acadians eventually became known as Cajun, and their culture spread across southern Louisiana, from the marshland to the prairies beyond the delta. It's particularly associated with the swampy forests along the Atchafalaya River. By the early twentieth century, the largest village there, Bayou Chene, would have been akin to any rural town—frame houses and churches and a general store, plus, in this case, a saloon that doubled as the funeral home—but Bayou Chene had one key distinction: instead of roads, it lay along a network of bayous. The sheriff rarely visited, and for those who chose to float in houseboats, no property taxes were ever due.

Historians estimate that 150,000 people lived scattered across twenty wetland communities in the delta at the dawn of the twentieth century. In the freshwater Atchafalaya swamps, they enjoyed a full faunal calendar:

they caught rabbits and squirrels and snipe in winter, raccoons and opossums and turtles in spring, catfish and crabs all year round. In the brackish marshlands, muskrat trappers marked their claims with cane poles and dug ditches that allowed their boat to pass through. Historians describe their system as carefully designed: no one would cut a ditch through a high ridge, knowing it could change the hydrology. They built dams to keep salt water out.

The men in the city finally took note of the marshland in the 1920s, when a booming national economy, paired with a shift in high fashion, turned muskrat furs into a suddenly hot commodity. This launched a land rush. The trappers had grown accustomed to traveling wherever they wished, but now people were buying up plots and allowing entry only to those who paid a fee. Armed gunboats began to patrol the swamps; a machine-gun battle in 1926 left one man dead and eleven wounded. Still, even in this era, the marshland was seen as land that could be sacrificed for the sake of the city upstream. The Bohemia Spillway is a prime example. New Orleans began to pursue this flood-relief outlet in 1924, while people were living along the riverbank. They were evicted, and their properties were assessed at a fraction of their full value. Today, in the mud at the edge of the river, the detritus of the lives they gave up can be found: cracked old plates and cups, rusted nails, the cypress posts that served as the foundation for their homes.

Just a few months after the spillway was completed, the 1927 flood rolled downstream. The new outlet helped relieve the pressure on New Orleans, although not as much as city officials wanted. So another hole was blasted in the levee. The water rushed through the crevasse at Caernarvon, covering almost all of Plaquemines Parish's east bank. The muskrat hunters, working with state officials, anchored rafts to trees, which helped save the rodents from drowning; but trapped on top, exposed to levels of sunlight that they rarely experienced in their usual nocturnal lifestyles, many went blind. Others starved. It's likely that more than half of Louisiana's muskrats perished in the flood.

In Plaquemines Parish, at least, the damage was temporary. The post-flood engineering has changed the Atchafalaya swamps completely and permanently. This is one part of the delta where land loss is decidedly not a problem: the Atchafalaya, as a still-open distributary, has been slowly filling its basin with sediment ever since the river was born. The engineering has supercharged the process. Since the post office in Bayou Chene shuttered in 1952, the town has been buried under twelve feet of mud. The live oak trees—*chênes,* in French—have been killed off by the river's rise. Once upon a time, the Corps of Engineers figured all this new mud might please locals, since more mud meant more land for planting crops. But the swampers I've talked to, people who make their living pulling up crawfish, say the region was ruined.

Just a few years after the floods, a commodity far more valuable than fur was found in the swamps and marshland: oil. Surveyors trudged through the delta, sinking up to their necks, assessing the prospects of this land that was barely land. When corporations liked what they saw, they found a way to seize it. One Indigenous family was forced out when an oil company shot their cows, destroying their livelihood. Some residents, unable to read English, signed papers they believed affirmed their ownership; instead, they were quitting their claims. Elsewhere, the companies simply drilled without bothering to ask permission.

Now the wetlands have been punctured by tens of thousands of wells. The canals through the marsh span more than ten thousand miles in total, according to one estimate—enough to cross Louisiana from east to west forty times. With the land dissolving into the ocean, with hurricanes slamming ashore, the last remaining ethnic villages are being slowly drained of their inhabitants. The new marshland neighborhoods mostly consist of collections of luxurious hunting and fishing camps like Myrtle Grove, which stand above the water atop their stilts.

Amid this charged landscape, the CPRA's plan for diversions was an instant lightning rod. The early example built at Caernarvon—the same site where the levee was blasted open in 1927—did help revitalize the oyster industry, as planned. But it did so at the cost of several privately held oyster leases that were doused with too much fresh water. Locals sued.

They finally lost their case in the state supreme court in 2004, just years before the CPRA decided to build an even bigger diversion across the river at Myrtle Grove. Despite the local resistance, the CPRA's proposal only grew through years, until the project called for a maximum outflow of 75,000 cubic feet per second—nearly ten times more water than passes through the Caernarvon Diversion.

Acy Cooper, a third-generation shrimper, knows that if the marshland is not saved, there may not be another generation in the chain. That's why he supports other projects that the CPRA has launched—using mud dredged out of the Mississippi to build marsh, particularly. But he pointed to what happened in Mississippi in the 2019 flood as an example of what the diversion would do to his local estuary: the blast of fresh water will devastate the shrimp. They will survive farther out in the salty gulf, but that is too far for Cooper to travel, given the business's tight margins. Fuel prices always rise; the money he gets for shrimp does not. Cooper compared the sediment diversion to a gun held to his head. "Either let me die slowly and I can adapt, or you just pull the trigger and kill me now. That's the way I feel about it," he said. "If you pull the trigger now, I'm dead."

In 2021, just before we met, the Corps of Engineers released a draft environmental impact statement for the Mid-Barataria Sediment Diversion. It confirmed Cooper's worst fears. The addition of fresh water would have "major, permanent, adverse impacts on brown shrimp abundance," as the report put it. Oysters would suffer, too. In the final draft of the study, the Corps noted that no matter what we do, at some point the price of local shrimp—a key part of Louisiana's culinary culture and identity—is likely to increase, but the diversion would deliver that price leap decades earlier than if the project weren't built.

It would deliver further problems as well. Tidal flooding would increase near homes in Myrtle Grove and other marshland communities, while the canals that residents use to travel to their favored fishing sites would become plugged with mud. The environmental impact statement suggested that the return of fresh water to Barataria Bay would pose a severe threat to its two thousand dolphins, since fresh water causes skin lesions that can lead to infections. A federal commission of marine scientists worries that the local population might be wiped out entirely. Dolphins are a protected species—one that has struggled since the 2010 *Deepwater Horizon* oil spill—but the CPRA received a federal waiver exempting the sediment diversion from the relevant laws.

Cooper, like many residents I've met in Plaquemines Parish, believes the diversion is just a big and flashy undertaking, one being pursued mostly because it's lucrative for the people in charge. He's skeptical that it will succeed. "Mother Nature been changing the geography of this country for millions of years. You think man is going to step in here and change it?" Cooper said. "Are we that naïve? That the same son of a bitch that messed it up is going to come fix it?"

It's surprisingly hard to answer that question. Nothing in the river's history matches the particulars of the Mid-Barataria Sediment Diversion. There have been natural crevasses, like Mardi Gras Pass; there are freshwater diversions, as at Caernarvon. In both categories, results have been mixed: there are outlets that have dumped out enough mud to build new marsh, and outlets where the rush of water ripped through the existing grasses, hastening the destruction of the coast. The river has its own logic: its choice of where to crevasse is dictated by a complex interplay between the underlying geology and the current shape of the river. We're ignoring that logic and planting our artificial crevasse at a site that matches our own desires. The shoal that Richie Blink showed me is encouraging but hardly incontrovertible proof of the concept.

The officially sanctioned method for assessing the new sediment diversion's prospects is to build complicated computer models showing how the flow of water and mud are expected to change in the coming years. The Corps of Engineers' environmental impact statement suggested, based on such models, that if we do nothing, then over the next fifty years (its most distant forecast) we will lose 340,000 acres of wetlands in the region near the diversion. If we build this outlet, and it works as planned, we will lose 330,000 acres. It's a net loss either way, and the difference is a mass of mud as big as two good-size airports.

So what hope is there for the delta? The CPRA is already designing another large sediment diversion, to be located upstream and on the river's east bank, and it is considering where other, smaller diversions might be built, too. Their presence could solve multiple problems: new outlets upstream of New Orleans, modeling suggests, might reduce the need to send so much floodwater through the Bonnet Carré Spillway, preventing more oyster die-offs and algal blooms. Since the more northern diversions would empty into stands of cypress swamps, they'd also help process farm fertilizers before they're passed on to the Gulf of Mexico. Still, it must be said that most of the world's large river deltas are the product of a singular era. They emerged some 7,500 years ago, after the glaciers finished melting, as the rapid rise of the seas steadied, giving rivers a chance to extend into the ocean. The problem for the Mid-Barataria Sediment Diversion, and for the delta more generally, is that now, as the climate warms, the seas are rising again. Many hydrologists, even some who support the diversion, note that the natural rate of delta land-building—just a few square miles each year at any given river outlet—is far slower than the rate that land will be swallowed.

Ehab Meselhe, a professor of river engineering at Tulane University, led the team that built the computer models used by the Corps of Engineers. When I spoke to Meselhe, he sounded skeptical of the effort to unleash the river. This surprised me. But Meselhe indicated that's he's not a decision maker—just an engineer. "My task was [to] quantify how much land," he said of the model, "and that's my quantification." He was not equipped to decide whether Plaquemines Parish's shrimping industry was a fair trade for a few thousand acres of wetlands.

The diversion is a state project, to be owned and operated by Louisiana. So the federal government did not need to assess this trade, either. Its role was simple: the Corps of Engineers needed to check that the project would neither mar the navigation channel nor weaken the levees; it needed to complete the studies required by the National Environmental Policy Act and ensure that the diversion would not damage wetlands protected by the Clean Water Act. It did not need to endorse the project. So in 2022 the Corps, having answered those questions, granted the state its permits.

Late the next year, contractors began the long, slow process of pouring out the concrete, building the latest control structure. After forty years of discussion and debate, the river's next great renovation was finally underway.

~~~

The Mississippi River is now a geological artifact of the Anthropocene Epoch. Even if we halt all engineering and let the concrete crumble, the work of the engineers will remain inscribed on the Earth. We've added a new chapter to the alluvial chronicle, one that will remain visible to whoever updates Harold Fisk's maps. As we tumble further into this strange new era, an old quote from the counterculture leader Stewart Brand has been making the rounds: "We are as gods, so we might as well get good at it." Brand, who used these words in his 1969 *Whole Earth Catalog*, meant to encourage self-sufficiency—his catalog was a clearinghouse for hippie-era back-to-the-landers, a renewed effort to create the forgone empire for liberty. But these days, Brand's words have been dusted off to justify new technological interventions. Since our influence is everywhere now, "nature" no longer really exists. If our control structures cause problems, we can just build more control structures to fix them. A pumping station can rewater a drainage district. A sediment diversion can release mud that's been trapped by levees.

This philosophy leaves me uneasy, with its you-broke-it-you-bought-it logic. If my time on the river has convinced me of one thing, it's that we do not make very good gods.

The river is the true god here—an unappeasable god, as the planter William Alexander Percy once remarked. Perhaps we need to acknowledge the power of this god, its right to build a landscape and then erase it again, its desire—and need—for endless change. River engineering, as hydrologists increasingly acknowledge, is always experimental, and the experiments don't often yield great results. "There aren't many engineering examples in history where we interfere with the natural system and made it better," Meselhe told me. "The less we tinker around with nature,
~~~

the better." When people discuss subsidence, he noted, they tend to express the problem in terms of how much land has been lost since some date in the past. The implicit suggestion is that we should try to restore the delta of that era. "I don't know if that is—not only if it is feasible—[but] if that's even the right way to look at it," he said. "Because the coast has always been meant to be a dynamic system."

What long allowed the delta to survive was the river's habit of leaping across its coastline in hundred-mile strides—finding different routes, building anew. Now we're granting it one artificial crevasse. It's an expensive crevasse, too. The projected $3 billion price tag for the diversion will mostly go to our efforts to control the water. A concrete trough, two miles long, will carry the flow through the riverside ridge, preventing erosion; as at Old River, steel gates will ensure that only the desired amount of water is released. These structures should assuage some fears, but they're also a reminder of how much of the river's power we plan to deny.

There's a way of restoring the delta that's far simpler than building endless diversions. We could accede to the river's logic. Reopen the old distributaries, that is, tear out the Old River Control Structure and let the river do what it does. Beginning in the 1950s, once the Atchafalaya's lakes were mostly filled with mud, a new delta began forming at the distributary's two mouths—a signal that the one remaining outlet carries sufficient sand to build new marshes. In the wake of an avulsion, the mad rush of water would create even more land. Tear down the levees lining the Atchafalaya Spillway, and the mud and water could run down a branching network of bayous and spread across a wide expanse of Louisiana's coast. Of course, to release the river down the Atchafalaya would be akin to removing the levees in the alluvial valley: it would solve ecological problems, perhaps, but at terrifying cost. New Orleans would become a shell of its former self, the surrounding ports impossible to access. A mightier Atchafalaya would also wipe away the old Cajun heartland. The backswamps here, according to some estimates, would be buried under twenty feet of fresh silt.

Once, after a trek through those swamps, I met Roy Blanchard, who traces his Louisiana ancestry back eleven generations. He was born in 1941, just as the Atchafalaya guide levees came up. Roy took a boat to school, at least for the years that he attended. He quit by seventh grade, frustrated that they never let him speak his native French, and started fishing. He worked alone, usually, meeting his father back at the dock to unload the spoils. He loved the work, and loved the water. His wife Annie, too, knows the swamps intimately; she was born in a houseboat on Lake Rond, inside the spillway, in the late 1940s. Her family did not move to dry land until she was old enough to go to school. Now the couple lives in a trim white house that sits just outside the western guide levee. Roy built the house himself, with cypress he pulled from the swamps. He built the levee, too, or at least rebuilt it, earning extra money in the off-season by joining work crews engaged in the periodic task of upkeep and improvement.

"Do you eat frog legs?" Annie asked me when I arrived at their home. Roy had seen the frogs jumping in Red Eye Swamp when he was baiting trotlines. He'd told his son, who'd caught the frogs, then gifted them back to his parents. Now Annie was frying them in a cast-iron Dutch oven to be served atop rice and stewed white beans. I had become the latest link in this chain of Cajun hospitality.

While we ate, I asked the Blanchards if they thought the basin might be saved, or if it was just a place where people would fish for however many years remained before the swamps were gone, filled by the endless mud.

"That's about it," Roy said, adding that it was too late to rebuild what he knew. "It's more than sad," he mused. Then he noted that his father, Wilmer Blanchard, Sr., had been sad before him. *Son, you never saw beautiful country,* the older man, born in 1910, used to say. Wilmer had known the Atchafalaya swamps back in their golden days, before the spillway, before the mud—a place of narrow coulees and winding bayous and majestic oak trees. Eventually, Wilmer had refused to even cross over the levees with Roy. Back then, Roy had been incredulous. He was making good money catching crawfish. He was having fun. How could the place be ruined?

Roy leaned back in his chair, recalling his old way of thinking, and chuckled—tickled by it, apparently. He's come around to his father's view. Still, his laugh sounded less mournful than accepting. The mud will come, he knows, as it always has. That is the way of the delta, the way of the great river.

Roy and Annie once had a water well dug on their property. The drill turned up cypress shavings 180 feet down, he said—remnants of ancient forests, long ago buried. That's how much sediment has accumulated over the past few thousand years. "Just think of what's down there," Annie said. A dinosaur, maybe, she joked; a different landscape, certainly: "There's changes all the time. We're just here for a little while."

The Lemon Tree Mounds (Plaquemines Parish, Louisiana).

Epilogue

The Water of the Future

When Harold Fisk determined that the Mississippi River was a "poised stream," one that had sustained an equilibrium in its length and depth for centuries, he was looking back, not forward. He hardly even considered the state of the river of his own era. He made only passing reference to Ferguson's cutoffs. He seemed unaware that as farmland upriver was being cleared, the amount of water pouring downstream was steadily increasing. It's unlikely he thought much about climate change, that within a few generations the seas might begin rising. But hither they come.

Already, these changes are rewriting the geology of the Birdfoot Delta. Under the pressure of so many recent floods, the unprotected riverbanks below the Bohemia Spillway have failed repeatedly, carving not just Mardi Gras Pass but other new outlets, too. Today, only a fifth of the water that runs past New Orleans reaches the three passes that are officially considered the river's mouth. One of the new exits—ripped open by the long 2019 flood—is already carrying more water than the engineers plan to grant to the Mid-Barataria Sediment Diversion. Less than two miles long, this pass is by volume one of the largest rivers in the country.

Various scientists and fishermen and public officials debated what to call this new channel. Eventually, the consensus settled on Neptune Pass. I rather prefer the suggestion offered by Richie Blink, my tour guide through the delta: Avulsion Pass, in recognition of the fact that the river here is straining toward a new route. As sea level rises, the Mississippi will slow in the Birdfoot Delta, which means it will deposit more mud in the riverbed. That will give the river even further reason to turn. Relatively speaking—compared to the shift through Old River—this would be a minor jump. Still, it's one more crack in the system.

Officials from the CPRA are excited to see new marshland growing at the end of Neptune Pass: a pretty swamp-meadow filled with cattail and Dutch potato, even a few young willow trees, in a place that just recently was open water. The Corps of Engineers, meanwhile, has already installed rocks along the riverbottom at the gap, preventing further erosion. The agency's surveys found shoals blocking the riverbed nearby and such a strong surge of water that boats were getting sucked into the pass. The plan is to mostly shutter the new outlet, though the Corps is studying whether it's possible to still allow sediment to be carried out into the marsh.

Old River, too, remains precarious. In 2018 a sand boil opened in the levee a few miles north of the complex of gates; a levee failure here would allow the water to sidestep the Old River Control Structure entirely, forging a new route into the Atchafalaya Basin. The army's geotechnical engineers began contemplating what they might do if the levee was blown away. Perhaps they'd fill the gap with rocks and sandbags; or maybe they'd open the dams upstream on the Red River, which feeds into the Atchafalaya, so that the force of the water could act as a battering ram and push back the Mississippi's flow. ("Army Corps, we can.") The mud that was carved off Ferguson's knickpoints, meanwhile, is still accumulating just downstream of Old River, increasing the chance of avulsion. And the Atchafalaya is still carving itself a bigger and better channel—and, according to models, will continue to do so for the next hundred-plus years, even with the control structure in place. Despite the engineering, researchers have concluded, the "the geologic evolution toward capture gradually continues."

In 2023, the Corps of Engineers launched what it calls a "mega-study": a five-year, $25 million analysis of the southern river and its infrastructure. The Lower Mississippi River Comprehensive Management Study will almost certainly dwarf Andrew Humphreys's old report; back then, the federal government was studying just flood control, even though the issue officially lay outside its responsibilities. Now the Corps is required to supply flood protection—and to sustain the river's many other "purposes," too. The mega-study will consider hydropower, ecosystems, and even river recreation.

"We've got a lot going on," said Ricky Boyett, the chief of public affairs for the agency's New Orleans District, which is leading the study. "It's all making sure we can pass what we're facing now—but also that we have a system that can continue to manage this water in the future."

The study will examine the impacts of new diversions, and consider what might happen if we adjust the amount of water sent through the existing control structures. To the north, in the alluvial valley, the Corps will have to assess the health of the MR&T system levees, to determine whether they're prepared to hold back the surge of water that may arrive in the next few years. But the study will have to reckon, too, with the fact that, after more than eighty years, the plans that the Corps laid in 1941 are still not complete.

As the MR&T Project came up, the Yazoo Basin became completely ringed in levees and bluffs. Now any rain that falls across a four-thousand-square-mile expanse has just one outlet: a set of four gates, at the southern edge of Issaquena County, that open onto the Yazoo River, just upstream of its mouth. When the Mississippi rises into floods, the gates must be kept shut to keep the big river's backwash from pushing through. But that causes a different sort of flood problem: when the gates are closed, the local rain has no way to drain.

The final plans for the MR&T Project called for a pump that could carry the water over the levee into the Yazoo River. But the backwater levee and the pumps both fell victim to the standard delays—too many projects, not enough money. Then in a rare case of restraint, a Corps analysis suggested that given the marginal state of the nearby farmland, the whole project wasn't worth the cost.

The rise of soybeans and the ensuing rush of land-clearing changed the calculus in the 1970s, and the levee came up. By then, though, environmentalists had decided that the pumps were likely to damage the last remaining stands of swamps within the Yazoo Basin. A conservative movement critical of pork barrel spending complained about the expense, too—many millions of tax dollars to help just a few local farmers. In 2008, after years of furious battles, the Environmental Protection Agency stepped in to kill the project. Today the "Yazoo Pumps," as they've become known, are the last substantial component of the MR&T Project that remains unbuilt.

From 2001 through 2023, the southern Yazoo Basin experienced some amount of backwater flooding in all but four years. But the worst came in 2019, the year of big rain. Local runoff, trapped by the closed gates, grew into a pool that reached forty miles north; it spanned, at its widest, twenty-eight miles from east to west. In June a man and a pregnant woman turned the wrong way while driving on a highway; their car slipped into the water, and both died. Because this inland sea included 230,000 acres of farmland, the economic tragedy was more widespread. "Ain't no money," one farmworker told Mississippi Public Broadcasting. "So we had to cut back on grocery shopping 'cause we had the light bills and the cable bills still coming in." The water, thick with sewage and motor oil, finally began to subside in late July. When residents returned, they found dead fish trapped in treetops; the ground was littered with the skeletal remains of drowned raccoons and deer. One Mississippi Levee Board commissioner came upon a "tree of death"—filled with hundreds of vultures, surrounded by hundreds of carcasses. "The things I saw I never want to see again," he later said. But the next year the region flooded once more. Bitter over the lack of media coverage, which focused on the struggles along the Missouri, many locals took to calling this the "Forgotten Flood."

The missing pumps, meanwhile, they described as a promise that the nation failed to keep. This fury registered, and federal officials descended on the region to hear locals' complaints. In the summer of 2023, as I was finishing this book, the Corps of Engineers announced a new proposal: in exchange for a bigger pump that could move water faster, they'd let the water rise a few feet higher before turning on the pumps. The engineers launched the required environmental study of this new proposal, hoping the raised threshold might lessen the impact on local swamps.

The pumps were initially described as an assist to the area's agricultural economy. Eventually, Mississippi politicians began to talk of their "human side," emphasizing that, as one senator claimed in the early 2000s, they would keep floodwaters out of a thousand homes. Both the Corps of Engineers and the Mississippi Levee Board, which serves as the local partner sponsoring this project, now describe the pumps as a matter of environmental justice, since many of those homes are minority-occupied. One Black man I met in Issaquena County, Anderson Jones, told me that when the floods began to rise, he sent his family to higher ground, but he stayed put himself, killing snakes, building a wall of sandbags. At night he sat alone in the darkness, miles from the nearest human being, listening to the bullfrogs and alligators groan. He lost his battle on May 19, 2019, at two a.m., when he awoke to find the water pouring in through his sewage lines. He cut the power and fled. "We didn't get a chance to save nothing," he told me. Even four years after the flood, he'd not managed to piece together the money to kill the black mold that had sprouted.

Many major environmental groups remain opposed to the pumps. "There's nothing new here," Louie Miller, director of the state's Sierra Club chapter, told a local reporter, describing the Corps' updated proposal. "They just put some lipstick on the pig." These groups maintain that the best future for the Yazoo backwaters might be to give up on the pumps and rebrand the region. There are still "hemispherically significant" wetlands here, after all. If environmental justice is truly the goal, then the money should go directly to suffering residents. Existing federal programs can pay locals to elevate homes or perhaps even to abandon their repeatedly flooded properties and find somewhere better to live.

Locals object, correctly, that these programs are slow-moving and often underfunded. And as compelling as it is to imagine the wetlands returning here, I can't help but notice a contrast: To the south, in the delta, the community leaders who have fought the plastic plants and refused to leave their homes have been praised by environmental groups for their resistance. In the Yazoo backwaters, meanwhile, where Black-held land might be turned into a nature park, they are told to pick up and move on.

To construct the Yazoo Pumps would cost, as of the latest estimate, $400 million—more than the Corps of Engineers spends on the entire MR&T system in some years. Divide that amount across a thousand flooded homes, and you could give each owner $400,000. Issaquena County, the site of the gates, is one of the poorest counties in the country; the median household income is $17,000. Four hundred thousand dollars could change lives. But simple division is a skewed way to calculate the payout. In a 2007 analysis, the Corps found that only 10 percent of the pumps' economic benefits would go to nonagricultural "structures." More than 75 percent would accrue to the farming industry. These figures are unlikely to change significantly in the newly proposed design. To call these pumps a tool of justice strikes me as, if not disingenuous, then at least grossly calculated.

The alluvial valley's planters have always been masterful at steering government assistance to their advantage. First they convinced the government to fund levees and drainage canals. Then they used New Deal programs to supply themselves with tractors and chemicals. Among the pumps' loudest proponents is the Delta Council, which spent most of the last century opposing federal welfare programs meant to assist suffering local populations. "Where's their concern been all along?" asks James Cobb, a historian who has studied the region's economy. "Their concern would appear to be very selective to me." Ray Mosby, the editor of a local newspaper, was more cutting. "They'll use 'em in a heartbeat," he told me. "They'll put them out there like red badges of courage."

I find the fight over the pumps curious: it's not as if keeping the pumps unbuilt will somehow restore the lost ecology of the alluvial valley. (To do that, we'd have to, among other unlikely steps, tear down some levees.) Nor will the installation of the pumps somehow make the southern Yazoo

Basin into a land of milk and honey. This expanse will remain what it has always been: a marginal bit of farmland. Yet three separate presidential administrations have intervened through the years to kill or to save these pumps. Newspapers across the country have covered the controversy. The pumps have become a symbol, I think—a memorandum on this American river. Do we want to keep doing things the way we've been doing them? Or is it time for something new?

I fear, though, that the smallness of the question—pumps or no pumps—leaves out too much. I'm less worried about the damage the pumps might do than the damage they will not fix. There are bigger disasters in the alluvial valley than flooding, after all: poverty, inequality, graft. We sometimes call the floods an act of god, but they're not really. They're just one more crisis delivered upon this place by our interventions. If the government can build a pump to stop the flooding we've caused, why not clean up the rest of our messes, too?

A few years ago I caught a ride in an oyster boat skimming atop the ghost of the marshland in Barataria Bay. We were headed toward the site of a mound complex, a place where the river's former inhabitants had built three earthworks. Or maybe four, it's hard to tell now.

Once these mounds rose atop a bayou-side ridge. That strip of land is long gone, sunk into the ocean. The bayou has become an invisible line, a trench at the bottom of this patch of salty water. Just one mound from the complex remains, barely cracking the surface of the gulf. Had I arrived alone, I would have noticed nothing extraordinary, just a few trees protruding from a small island of marsh grass. But Rosina Philippe, an elder from the Grand Bayou Atakapa-Ishak/Chawasha Tribe, pointed out the slight rise of the mound, its core now exposed to the assault of the waves. Her father used to stop here when he was fishing to make himself a glass of fresh-squeezed lemonade, she said. Just a generation ago, a citrus grove flourished atop the mound.

Philippe, along with the nonprofit Coalition to Restore Coastal Louisiana (CRCL), had recruited a group of volunteers to stack sacks of oyster shells in the water along the edge of the mound. The goal was to create a reef to dampen incoming waves, protecting what's left of this earthwork. CRCL has built other reefs, which they've found can cut erosion in half. Indeed, a few months after my visit, when Hurricane Ida tore through Louisiana, ripping out threads of marshland, the reef we visited remained intact. That won't be the case forever, given that this is the delta. The mound will keep sinking until it disappears beneath the waves. But Philippe indicated that the new oyster reef would remain, at least, a marker of this sacred place.

Philippe, like many members of her tribe, lives in a village in Plaquemines Parish that is accessible only by boat. She had recently learned that, at the tribe's request, the CPRA would help backfill local oil canals, preventing the further intrusion of salt and helping sustain what little scraps of land remain. "It's kind of late, but we'll take what we can get," she said. Her ancestors have known this landscape for millennia, she noted, and yet for so long no one bothered to ask what they knew or what they wanted.

Journalists sometimes invoke our work in the delta—from the gates at Old River to the sediment diversion now being constructed—as a microcosm of our fight against nature. But I'm not sure that what we're battling here is nature. Over the two centuries that engineers have been reworking this river, they've slowed some parts of the geology while pushing others into hyperdrive. The fight along the Mississippi River can't be against nature, since it's impossible to say what "nature" is here. It's always a fight among humans, people who can't agree what kind of world they want to build.

The prevailing yardstick to assess these projects has mostly been their economic impacts. The river's purpose, to the first Americans, was to be a highway; the floodplain's purpose was to be turned into property—though eventually, thankfully, a few of its loveliest slivers were granted reprieve. It's telling, I think, that the cost-benefit analysis now often suggests that we need to undo what we've built: that we punch holes in the delta levees to bring back the marshland and so protect our cities against hurricanes; that we pull back the levees upstream, decreasing the risk of flooding on our farms. Still, there is much that numbers will never capture. One report commissioned by the CPRA included a valuation of the marshland's neotropical bird population. The figure was calculated based on the millions of dollars tourists spend to watch their migration. It's a valiant effort to capture the worth of this place, perhaps, but to me it reflects flawed thinking.

I'm sometimes tempted to suggest we tear down the levees and give back the river its whole domain. We could let the southern floodplain revert to wilderness. We could make the entire valley a national park—a monument to itself, to the beauty of the world when it's left alone. But when I stop and think, I realize how foolish this is: not just an impossible dream but a wantonly destructive proposal. We'd lose the farms here—and they're essential, even if they should be better used. More than that, the resulting floods would wash away some of the nation's poorest communities. So how to move forward?

It's that question that had brought me here, to Barataria Bay, to this broken mound. The earthworks scattered along the river are rooted in the swamps and their life-giving chaos; they reach toward the sky and its orderly march. To some modern minds, these layered worlds may sound mythical, but are they any more mythical than the Western idea that humans and nature are separate—that some parts of nature are so separate that they can be considered "wild"? Throughout this book, I've called the river a wild place, but these mounds contradict me. They offer a reminder that this river has never been alone. This is, after all, one of the youngest landscapes on the continent. Twelve thousand years ago, as the mud began to fill the watery basin, people were here already, fishing in the fresh swamps.

Perhaps this mound offers a different kind of monument, not to the beauty of empty nature, but to the possibility of a human connection. And not just a connection to nature. The construction of a mound could be, to borrow the words of the anthropologist Jay Miller, a "community-based action done in prayerful manner." The levee helped build an empire that, in the words of bluesman Charley Patton, made every day feel like murder, but the mounds are meant to serve everyone.

The earthworks did not arise in an Eden. It's impossible to live along the Mississippi River and decide that nature is gentle. The river is an unappeasable god, and to react to it with fear and awe is not wrong. So I do not see the earthworks as celebratory monuments. They're more like insurance, anchors amid the chaos. Perhaps what people learn after thousands of years of living along one of the world's great rivers is that change is inevitable, that chaos will come. That the only way to survive is to take care—of yourself and of everyone else, human and beyond.

Philippe told me her ancestors did not try to fight nature or conquer it. They accepted its floods. They accepted its mud, too—and saw it as a gift, a rich supply of new soils and nutrients that feed crops and shrimp and fish. "Our lives are possible because of all these other lives," she said as we cut across the former marshland, which now sparkled under the summer sunshine—nothing but open water. "Any one thing you take out, its absence will be known."

This strikes me as a lesson that has been too long ignored on this great river. It's not just land we've lost along this river. There's more than land that we need to restore.

Acknowledgments

I have been working on *The Great River* for nearly as long as I have considered myself a writer. Like the Mississippi River itself, this project offered a lesson in interconnectedness: while it's my name on the cover, the book could not exist without the kind assistance of so many others—some of whom I've inevitably but regrettably left out.

To start at the beginning: This project was born as a magazine story about John Ruskey, who has been my stalwart guide to the river ever since. His wisdom, his kindness, and the example he has set are inspiring. Many other paddlers are due thanks, too, including Chris "Magique" Battaglia, Mike "Big Muddy" Clark, Adam Elliot, Birney Imes, Layne Logue, Andy McClean, and Mark "River" Peoples. Photographer Rory Doyle has accompanied me on many of my river adventures, and his keen eye has shaped my vision of this waterway.

My agent, Barney Karpfinger, recognized a kernel of promise in my notes from these travels, and gently and deftly coached me on how to think and talk about the project. Sam Chidley, too, provided valuable notes and support. They shepherded me to my editor, Matt Weiland, who helped me carve a watershed worth of reporting down to manageable form; his enthusiasm has been infectious. Huneeya Siddiqui has endured my perfectionist tinkering and ably steered the ship through the treacherous shoals of production. My wife, Liz, in addition to so much else, created a stunning (and extensive) portfolio of maps. It's been a dream team.

Funding for my Mississippi River reporting has come from the Mississippi Delta National Heritage, Delta State University (through the Teach For America Fellowship), the University of California, Berkeley, School of Journalism's 11th Hour Food and Farming Fellowship, and the Institute for Journalism and Natural Resources. Individual contributions helped keep the project alive in its lean early years, with especially generous gifts from Shannon Brady, Wes Care, Brian Doigan, Rory Doyle, Jackson Edwards, Jossi Fritz-Mauer, Shelby Goss, JB Haglund, Bernie Jones, Ron Nurnberg, Nate Rosenthal, David Schlesinger, Ethan Smith, Gretchen Upholt, Mary Lee Morrison and William Upholt, and Diane Weinholtz.

Peter Buckley was a tireless research assistant throughout reporting; Gwen Dilworth braved the task of fact-checking these pages on a short schedule and did a remarkable job. An insightful band of readers helped me hone early drafts: Jacob Carroll, Jason Christian, Philip Kiefer (who provided an essential fact-check assist), Amy Lin, and David Schlesinger, as well as my advisors at DSU, David Baylis and Chuck Westmoreland. Adam Crosson was an essential link to libraries in the strange era of the coronavirus. Matty Bengloff has been a relentless cheerleader as I've stumbled across the finish line.

In addition to the scientists, engineers, and river rats mentioned in the text—whose willingness to share their knowledge was humbling—I was honored to discuss and explore the lower watershed with many whose names had to be skipped. In the Atchafalaya Basin: Thomas Edgar Ashley, Joe Baustian, Rien Fertel, Monica Fisher, Scott Green, Greg Guirard, Jody Meche, Gerrard Perrone, Rudy Sparks, Jo Vidrine, and Dean Wilson. In the alluvial valley: Ed Adcock, Scott Barretta, Hank Burdine, Tom Burnham, Hannah Conway, Jessica Crawford, John Crilly and Lanchi Luu, Greg Dycus, Terry Fuller, Diana Greenlee, Warren Hines, Matthew Horner, Alan Huffman, Sammy King, Sam Muñoz, Cliff Ochs, Peter Rost, Mike Sullivan, Tim Sullivan, and David Wildy. In the delta: Mike Beck, Rich Campanella, Craig Colten, Hali Dardar, Jeffery Darensbourg, Macon

Fry, James Karst and the staff of the CRCL, plus various staff members of the CPRA, Alex Kolker, Brian Ostahowski, and June Provost. On the Middle Mississippi: Ray Cole, Bob Criss, Jeff Denny, and Lori Smith. On the Missouri: Leo Ettleman, Regan Griffin, Ron Schneider, David Sobczyk, and Seth Wright. Robbie Ethridge, Robert Gudmestad, Chris Morris, and Thomas Ruys Smith lent their knowledge of the river's history; archeologists Gayle Fritz, Megan Kassabaum, and Kenneth Sassaman helped direct me through the overwhelming amounts of research on the prehistoric river, and Jay Miller offered wisdom, anthropological and beyond. Randy Cox double-checked my attempts to describe geology. Paul and Libby Hartfield, in addition to their river knowledge, provided a retreat amid a deadline crunch, and simply good friendship.

The Upper Mississippi remains for me a foreign territory. John Anderson, Frank Bibeau, Deb Calhoun, Greg Genz, Rylee Hince, Dean Klinkenberg, Steve Kujak, Hokan Miller, Pat Nunnally, and the staff of the Minneapolis Parks Foundation provided knowledge and access. Michael Anderson and Terry Larson were knowledgeable guides to gorgeous water. Katie Kelly and Ryan White, thank you for a fine crash pad. To Bill Waldschmidt: thanks for the meal and conversation and the gift—though I camped out anyway.

Finally, thanks to my family—Mom, Gretchen, and especially Liz, my wife and best friend. Without your patience, love, and support, this book, a lifelong dream, could not exist.

Notes

Since the founding of the United States of America, that last word—*America*—has been deployed as shorthand for the rather cumbersome full name. It's a problematic abbreviation, since the same word appears in the names of two continents. (Let's leave aside the logic by which so much of the world's landmass got named for the Italian explorer Amerigo Vespucci.) For the first generations of U.S. citizens, the words *America* and *American* had specific and narrow meanings: they referred to a union of white people, mostly descendants of British settlers, who believed they were creating a new kind of nation and government. Throughout this book, when I use these words, I mean them in this limited sense, acknowledging without endorsing an out-of-date idea. My occasional use of the words *slave* and *Indian* follows the same logic. When referring to Indigenous people in my own voice, I have wherever possible tried to use the name a person would use to identify his or her own people.

Introduction: A Most Magnificent Spectacle

1 ***Misi-ziibi*:** Ojibwe scholar Anton Treuer told me that *Misi-ziibi* translates to "long river," though he noted that there are several similar phrases used in various Ojibwe dialects. *The Ojibwe People's Dictionary*, a project of the University of Minnesota, links the name Mississippi to the word *Gichi-ziibi*, which is defined as "a big river." Elsewhere in the dictionary, *gichi-* is translated as both "big" and "great." (*Misi-* does not appear in the dictionary.) *Ojibwe People's Dictionary*, https://ojibwe.lib.umn.edu/main-entry/gichi-ziibi-ni.

2 **More than a hundred thousand named waterways:** Ari Kelman, *A River and Its*

City: The Nature of Landscape in New Orleans (Los Angeles: University of California Press, 2006), 1–2. According to the CIA's *World Factbook,* the Mississippi is the world's fourth-largest watershed, after the Amazon, the Congo, and the Nile. Official sources vary in their count of states in the watershed; here I include South Carolina, whose northwestern border mostly tracks the edge of the watershed but, perhaps because of slip-ups by surveyors, includes a few tiny bits of land that drain into the Tennessee River.

2 **140 cubic miles of water:** Dongho Lee and Ján Veizer, "Water and Carbon Cycles in the Mississippi River Basin: Potential Implications for the Northern Hemisphere Residual Terrestrial Sink," *Global Biogeochemical Cycles* 17, no. 2 (2003): 6.

3 **"turbid and boiling torrent":** A. A. Humphreys and H. L. Abbot, *Report Upon the Physics and Hydraulics of the Mississippi River* (Philadelphia: J.B. Lippincott, 1861), 28.

3 **"It was a most magnificent spectacle":** Garcilaso del Vega, *The Florida of the Inca,* trans. Jeannette Johnson Varner and John Grier Varner (Austin: University of Texas Press, 1951), 554.

3 **more than fifteen thousand square miles:** John A. Baker et al., "Aquatic Habitats and Fish Communities in the Lower Mississippi River," *Aquatic Sciences* 3, no. 4 (1991): 340.

3 **at least 24 million acres:** Ann Vileisis, *Discovering the Unknown Landscape: A History of America's Wetlands* (Washington, D.C.: Island Press, 1997): 22–24.

4 **"The men of the Mississippi Valley":** Frederick Jackson Turner, *The Frontier in American History* (1893; reprint New York: Henry Holt, 1921), 185. Turner's thesis that the frontier helped form American identity has been widely influential—not just among academics but in pop culture, too, informing the genre of "Western" novels and films. It's also been criticized by scholars who point out that the idea necessarily excludes the continent's Indigenous population from being considered American. (Besides race, Turner has also been criticized for failing to consider class and gender sufficiently.)

5 **"the shifting unappeasable god":** William Alexander Percy, *Lanterns on the Levee: Recollections of a Planter's Son* (Baton Rouge: Louisiana State University Press, 2011), 4.

5 **first levee:** The historical records of this first levee are murky. The best contemporary source documenting its beginnings comes from a letter, written in 1720, that indicates that enslaved Africans were at work. Richard Campanella, a prominent New Orleans historian, told me he figures construction began in 1719, soon after the flood receded, and that the initial round of work was completed at some point in 1720. Campanella also provided the details of this first levee included in my description here.

5 **the Île d'Orléans:** Richard Campanella, *Draining New Orleans: The 300-Year Quest to Dewater the Crescent City* (Baton Rouge: Louisiana State University Press, 2023), 19.

5 **"arguably the most successful":** U.S. Army Corps of Engineers, "MRC History," https://www.mvd.usace.army.mil/About/Mississippi-River-Commission-MRC/History.

5 **There are no dams:** For a good overview of the engineering across the watershed,

see Jason S. Alexander et al., *A Brief History and Summary of the Effects of Engineering and Dams on the Mississippi River System and Delta,* Circular no. 1375 (Reston, Va.: U.S. Geological Survey, 2012).

6 **The world's longest levee:** Paul Hudson, *Flooding and Management of Large Fluvial Lowlands: A Global Environmental Perspective* (New York: Cambridge University Press, 2021), 168.

6 **forty feet tall:** "Levee History," Yazoo-Mississippi Levee District, https://www.leveeboard.org/history12.html.

7 **"river of desolation":** Unless otherwise noted, the quotations from Mark Twain are drawn from *Life on the Mississippi* (1883), at https://gutenberg.org/files/245/245-h/245-h.htm.

9 **which supply a billion pounds:** "What's at Stake," Restore the Mississippi River Delta, https://mississippiriverdelta.org/whats-at-stake/; Richard Cody, *2020 Fisheries of the United States* (National Oceanic and Atmospheric Administration, 2022), 10.

9 **eight hundred thousand square feet:** Brady R. Couvillion et al., *Land Area Change in Coastal Louisiana, 1932 to 2016* (Reston, Va.: U.S. Geological Survey, 2017), 1. This rate is much slower than it was a few decades back, when daily losses averaged 2.4 million square feet.

10 **hit its "ten-year flood stage":** Michael Grunwald, "A Decade After Katrina, Are America's Flood Estimates Dangerously Wrong?" *Politico,* August 19, 2015.

10 **"Animals and Beggars":** Jorge Luis Borges, *The Aleph and Other Stories* (New York: Penguin, 2000), 181.

1. Searching for a River

Two books provide a thorough overview of Soto and his wandering: Charles Hudson, *Knights of Spain, Warriors of the Sun: Hernando de Soto and the South's Ancient Chiefdoms* (Athens: University of Georgia Press, 1997), and Robbie Ethridge, *From Chicaza to Chickasaw: The European Invasion and the Transformation of the Mississippian World, 1540–1715* (Chapel Hill: University of North Carolina Press, 2010). The classic work on La Salle—dated now but still a tremendous piece of scholarship—is Francis Parkman, *France and England in North America,* vol. 3, *La Salle and the Discovery of the Great West* (Boston: Little Brown, 1908). My version of La Salle's voyages also draws from Christopher Morris, *The Big Muddy: An Environmental History of the Mississippi and Its Peoples from Hernando de Soto to Hurricane Katrina* (New York: Oxford University Press, 2012); Lawrence Powell, *The Accidental City: Improvising New Orleans* (Cambridge, Mass.: Harvard University Press, 2012); and Paul Schneider, *Old Man River: The Mississippi in North American History* (New York: Picador, 2014).

15 **"wild land":** Quoted in Dan Flores, *Jefferson and Southwestern Exploration: The Freeman and Custis Accounts of the Red River Expedition of 1806* (Norman: University of Oklahoma Press, 1984), 3.

16 **"humbler growth":** Quoted in John O. Anfinson, *The River We Have Wrought: A*

History of the Upper Mississippi (Minneapolis: University of Minnesota Press, 2003), 12.

17 **"one of the most tranquil":** Henry Rowe Schoolcraft, *Summary Narrative of an Exploratory Expedition to the Sources of the Mississippi River in 1820* (Philadelphia: Lippincott, Grambo & Co., 1855), 242.

17 **"earliest explorers be not robbed":** Quoted in John H. Baker, "The Sources of the Mississippi," in *Collections of the Minnesota Historical Society* (St. Paul: Pioneer Press, 1894), 6:27.

17 **"from the earliest times":** Quoted in Rich Heyman, "Locating the Mississippi: Landscape, Nature, and National Territoriality at the Mississippi Headwaters," *American Quarterly* 62, no. 2 (2010): 323.

18 **"a sight that is not becoming":** Heyman, "Locating the Mississippi," 309.

18 **600 million years ago:** For an excellent introduction to the geology of the lower river, see Randel Tom Cox, "A Geologist's Perspective on the Mississippi Delta," in *Defining the Delta: Multidisciplinary Perspectives on the Lower Mississippi River Delta,* ed. Janelle Collins (Little Rock: University of Arkansas Press, 2015), 11–24.

19 **eighteen miles apart:** National Research Council, *The Missouri River Ecosystem: Exploring the Prospects for Recovery* (Washington, D.C.: National Academies Press, 2002), 56.

19 **spanned two miles:** Anfinson, *River We Have Wrought,* 12.

21 **"one unbroken slough":** Charles Dickens, *American Notes, and Reprinted Pieces* (London: Chapman & Hall, 1868), 105.

21 **ninety miles across:** Edward J. Brauer et al., *Geomorphology Study of the Middle Mississippi River* (U.S. Army Corps of Engineers, St. Louis District, 2005), 3.

21 **"land's slow alluvial chronicle":** William Faulkner, *Big Woods: The Hunting Stories* (New York: Vintage Books, 1994), 4.

21 **sixteen such cutoffs:** Brien Winkley, *Man-Made Cutoffs on the Lower Mississippi River, Conception, Construction, and River Response* (U.S. Army Corps of Engineers, Vicksburg District, 1977), 25.

22 **little more than fifteen feet above:** Roger T. Saucier, *Geomorphology and Quaternary Geologic History of the Lower Mississippi River* (Vicksburg: U.S. Army Engineer Waterways Experiment Station, 1994), 1:99.

22 **as much as a hundred feet deep:** Saucier, *Geomorphology,* 76.

22 **A surveyor:** The surveyor's notes are excerpted in Mikko Saikku, *This Delta, This Land: An Environmental History of the Yazoo-Mississippi Floodplain* (Athens: University of Georgia Press, 2005), 41.

23 **three bison napping:** Jeffery U. Darensbourg, "Hunting Memories of the Grass Things: An Indigenous Reflection on Bison in Louisiana," *Southern Cultures,* https://www.southerncultures.org/article/hunting-memories-of-the-grass-things/

24 **where rock appears:** Charles Ellet, Jr., *Report on the Overflows of the Delta of the Mississippi* (Washington, D.C.: A. Boyd Hamilton, 1842), 4.

24 **two to three square miles per year:** This is the rate of growth at a single major subdelta. See Elizabeth L. Chamberlain et al., "Anatomy of Mississippi Delta Growth and Its Implications for Coastal Restoration," *Science Advances* 4, no. 4 (2018): https://doi.org/10.1126/sciadv.aar4740.

24 **rarely tops ten feet:** Saucier, *Geomorphology,* 30.

27 **"A man standing on the shore":** Quoted in Edward Bourne, ed., *Narratives of the Career of Hernando de Soto,* trans. Buckingham Smith (New York: Allerton, 1922), 1:115.

30 **half-million:** These population estimates are from Robbie Etheridge, personal communication, 2023.

30 **"The collapse of the Mississippi":** Robbie Ethridge, "The Emergence of the Colonial South," in *Native American Adoption, Captivity, and Slavery in Changing Contexts,* ed. Max Carocci and Stephanie Pratt (New York: Palgrave Macmillan, 2012), 52.

31 **a hunter could receive:** Ethridge, "Emergence of Colonial South," 61.

31 **"Mississippian shatter zone":** Ethridge, *From Chicaza to Chickasaw,* 4.

31 **the entire nation would consist:** James F. Barnett, Jr., *The Natchez Indians: A History to 1735* (Jackson: University Press of Mississippi, 2007), 50.

31 **"quite useless":** Quoted in Powell, *Accidental City,* 20.

32 **"plunged into every kind":** Quoted in John Gilmary Shea, *Discovery and Exploration of the Mississippi Valley* (Clinton Hall, N.Y.: Redfield, 1852), 189.

32 **"fatal error":** Parkman, *France and England in North America,* 366.

33 **the Mississippi's delta is river-dominated:** James M. Coleman and Oscar K. Huh, *Major World Deltas: A Perspective from Space* (Baton Rouge: Coastal Studies Institute, Louisiana State University, n.d.), 8–10.

33 **failed to find the river:** Jack E. Davis, *The Gulf: The Making of an American Sea* (New York: Liveright, 2017), 35.

2. Tomahawk Claims

Daniel Usner, *Indians, Settlers, and Slaves in a Frontier Exchange Economy: The Lower Mississippi Valley Before 1783* (Chapel Hill: University of North Carolina Press, 1992), offers an excellent overview of the era considered in this chapter. Richard White's *The Roots of Dependency: Subsistence, Environment, and Social Change Among the Choctaws, Pawnees, and Navajos* (Lincoln: University of Nebraska Press, 1983) adds useful context about Choctaw culture in particular. For a good overview of the Natchez's fatal conflicts, see James F. Barnett, Jr., *The Natchez Indians: A History to 1735* (Jackson: University Press of Mississippi, 2007).

35 **"felt rather than seen":** Pekka Hämäläinen, *Lakota America* (New Haven, Conn.: Yale University Press, 2019), 40.

36 **The Ojibwe called these southern neighbors:** "The Illinois Identity," MuseumLink Illinois, Illinois State Museum, 2000, https://www.museum.state.il.us/muslink/nat_amer/post/htmls/il_id.html.

39 **"good site":** Quoted in Lawrence Powell, *The Accidental City: Improvising New Orleans* (Cambridge, Mass.: Harvard University Press, 2012), 11.

39 **"nothing more than":** Quoted in Richard Campanella, *Bienville's Dilemma: A Historical Geography of New Orleans* (Lafayette: Center for Louisiana Studies, 2008), 105.

39 **a half-dozen palm-roofed shacks:** The best description of this fort is Maurice Ries,

"The Mississippi Fort, Called Fort de la Boulaye," *Louisiana Historical Quarterly* 19, no. 4 (1936).

39 **"four months in the water":** This and the following quote appear in Christopher Morris, *The Big Muddy: An Environmental History of the Mississippi and Its Peoples from Hernando de Soto to Hurricane Katrina* (New York: Oxford University Press, 2012), 45.

40 **promised to deliver:** Richard Campanella, *Draining New Orleans: The 300-Year Quest to Dewater the Crescent City* (Baton Rouge: Louisiana State University Press, 2023), 15.

41 **"rather good road":** Quoted in Tristan R. Kidder, "Making the City Inevitable: Native Americans and the Geography of New Orleans," in *Transforming New Orleans and Its Environs*, ed. Craig Colten (Pittsburgh: University of Pittsburgh Press, 2000), 17.

41 **half the colony's population:** Usner, *Indians, Settlers, and Slaves*, 33.

45 **"overflowing Scum":** Quoted in Greg Grandin, *The End of the Myth: From the Frontier to the Border Wall in the Mind of America* (New York: Metropolitan Books, 2019), 17.

45 **"savage groves":** Quoted in Thomas Ruys Smith, *River of Dreams: Imagining in the Mississippi Before Mark Twain* (Baton Rouge: Louisiana State University Press, 2007), 38.

45 **"tomahawk claim":** William R. Nester, *George Rogers Clark* (Norman: University of Oklahoma Press, 2012), 17.

46 **the mouth of the Yazoo River:** The Yazoo was at the center of so many speculative dreams in this era. Twice in the late eighteenth century, the Georgia legislature sold vast tracts along the river to private companies, despite lacking standing to receive the land from its Choctaw and Chickasaw owners. The deals were rife with fraud. The land changed hands, increasing in value, despite the deals' illegitimacy. Ultimately, the federal government had to pay out millions to untangle the claims. See Jane Elsmere, "The Notorious Yazoo Land Fraud Case," *Georgia Historical Quarterly* 51, no. 4 (1967): 425–42.

46 **more than one hundred thousand people:** Malcolm Rohrbough, *Trans-Appalachian Frontier: People, Societies, and Institutions, 1775–1850* (Bloomington: Indiana University Press, 2008), 30.

47 **travelers were reporting:** Rohrbough, *Trans-Appalachian Frontier*, 52.

47 **The first recorded voyages:** Leland Baldwin, "The Rivers in the Early Development of Western Pennsylvania," *Western Pennsylvania Historical Magazine* 16, no. 2 (1933): 83.

47 **"The devil take it all":** Quoted in William E. Foley, *The Genesis of Missouri: From Wilderness Outpost to Statehood* (Columbia: University of Missouri Press, 1989), 79.

47 **"a streight of the sea":** Thomas Jefferson, "Report on Negotiations with Spain," March 18, 1792, at https://founders.archives.gov/documents/Jefferson/01-23-02-0259.

3. Cosmic River

Archaeologists have done essential work unwinding the story of the river's ancient earthworks, though they have also been prone to certain projections. Megan C. Kassabaum's *A*

History of Platform Mound Ceremonialism (Gainesville: University of Florida Press, 2019) offers a good corrective to some of the old misinterpretations and a useful summary of the history of mound building. My thoughts on earthworks were particularly influenced by Kenneth E. Sassaman and Asa R. Randall, "Cosmic Abandonment: How Detaching from Place Was Requisite to World Renewal in the Ancient American Southeast," in *Detachment from Place: Beyond an Archaeology of Settlement Abandonment*, ed. Maxime Lamoureux-St-Hilaire and Scott A. Macrae (Denver: University Press of Colorado, 2020). For Cahokia specifically, essential sources include Gayle Fritz, *Feeding Cahokia: Early Agriculture in the North American Heartland* (Tuscaloosa: University of Alabama Press, 2019), and Timothy Pauketat, *Cahokia: Ancient America's Great City on the Mississippi* (New York: Penguin Books, 2010).

50 **"a temple for the adoration":** Dunbar's notes on the site appear in Terry Berry et al., eds., *The Forgotten Expedition, 1804–1805: The Louisiana Purchase Journals of Dunbar and Hunter* (Baton Rouge: Louisiana State University Press, 2006), 190.

50 **"grounded *earth*":** Chadwick Allen, *Earthworks Rising: Mound Building in Native Literature and Arts* (Minneapolis: University of Minnesota Press, 2022), 10.

51 **"the capital of an extensive":** Winslow M. Walker, "The Troyville Mounds, Catahoula Parish, La.," *Bureau of American Ethnology Bulletin* 113 (1936): 4.

51 **"finally became so unreasonable":** Walker, "Troyville Mounds," 2.

52 **more than fourteen thousand years ago:** Jessi J. Halligan et al., "Pre-Clovis Occupation 14,550 Years Ago at the Page-Ladson Site, Florida, and the Peopling of the Americas," *Science Advances* 2, no. 5 (2016): https://www.science.org/doi/10.1126/sciadv.1600375.

52 **oldest formal open-air cemetery:** Kenneth Sassaman, personal communication, 2021.

52 **Watson Brake:** Joe W. Saunders et al., "Watson Brake, a Middle Archaic Mound Complex in Northeast Louisiana," *Society for American Archaeology* 70, no. 4 (2005): 631–68.

53 **One potential culprit:** Tristam R. Kidder et al., "Basin-Scale Reconstruction of the Geological Context of Human Settlement: An Example from the Lower Mississippi Valley, USA," *Quaternary Science Reviews* 27 (2008): 1255–70.

54 **seemed committed:** Pauketat, *Cahokia*, 17.

57 **According to archaeologists' calculations:** Annalee Newitz, *Four Lost Cities: A Secret History of the Urban Age* (New York: W.W. Norton, 2021), 208.

57 **cosmogram:** Caitlin G. Rankin et al., "Evaluating Narratives of Ecocide with the Stratigraphic Record at Cahokia Mounds State Historic Site, Illinois, USA," *Geoarchaeology* 36, no. 3 (2021): 371.

58 **"The Vacant Quarter":** Robbie Ethridge, *From Chicaza to Chickasaw: The European Invasion and the Transformation of the Mississippian World, 1540–1715* (Chapel Hill: University of North Carolina Press, 2010), 123.

59 **a false presumption:** The best rebuttals of the various ecocide arguments appear in Rankin et al., "Evaluating Narratives of Ecocide," and Fritz, *Feeding Cahokia*.

60 **as tall as Monks Mound:** Molly Langmuir, "What, a Dump?" *Riverfront Times* (St. Louis), November 8, 2006.

61 **"Indigenous knowledge":** Allen, *Earthworks Rising*, 62. For an immersive imaginative rendering of the human experience at Cahokia, I highly recommend Phillip Carroll Morgan's novel *Anompolichi: The Wordmaster* (Ada, Okla.: White Dog Press, 2014).

61 **these realms can overlap:** LeAnne Howe, "Embodied Tribalography: Mound Building, Ball Games, and Native Endurance in the Southeast," *Studies in American Indian Literatures* 26, no. 2 (2014): 84.

61 **"a kind of theatrical performance":** LeAnne Howe, "Life in a 21st Mound City," in *The World of Indigenous North America*, ed. Robert Allen Warrior (New York: Routledge/Taylor & Francis Group, 2015), 7.

61 **"geographic onomatopoeia":** Quoted in Sassaman and Randall, "Cosmic Abandonment," 70.

61 **cosmic river:** Sassaman and Randall, "Cosmic Abandonment," 70.

63 **"Rapid and catastrophic flooding":** Tristram R. Kidder, "Climate Change and the Archaic to Woodland Transition (3000–2500 Cal B.P.) in the Mississippi River Basin," *American Antiquity* 71, no. 2 (2006): 218.

64 **Some supposed:** For an overview of settler views of the earthworks, see Gordon Sayre, "The Mound Builders and the Imagination of American Antiquity in Jefferson, Bartram, and Chateaubriand," *Early American Literature* 33, no. 3 (1998): 225–49; and Angela Miller, "The Soil of an Unknown America: New World Lost Empires and the Debate over Cultural Origins," *American Art* 8, nos. 3–4 (1994): 8–27.

65 **"ignored or maybe even":** Mark A. Rees and Rebecca Saunders, "Mississippian Culture," *64 Parishes*, updated June 30, 2021, at https://64parishes.org/entry/mississippian-culture; and Mark A. Rees and Rebecca Saunders, "Plaquemine Culture," *64 Parishes*, updated August 24, 2022, at https://64parishes.org/entry/plaquemine-culture.

65 **Grandmother:** My discussion of Grandmother is based on conversations with Gayle Fritz. Many of these details can be found in her *Feeding Cahokia* and in Natalie G. Mueller and Gayle J. Fritz, "Women as Symbols and Actors in the Mississippi Valley," in *Native American Landscapes: An Engendered Perspective*, ed. Cheryl Claasen (Knoxville: University of Tennessee Press, 2016), 114–17.

65 **Indigenous oral histories:** For the Osage stories, see Meghan E. Buchanan, "Diasporic Longings? Cahokia, Common Field, and Nostalgic Orientations," *Journal of Archaeological Method and Theory* 27 (2020): 72–89. Allen details the Cherokee stories in *Earthworks Rising*, 303–6.

66 **the two honored bodies:** Thomas E. Emerson et al., "Paradigms Lost: Reconfiguring Cahokia's Mound 72 Beaded Burial," *American Antiquity* 81, no. 3 (2016): 405–25.

66 **to honor ancestors at large:** Robin A. Beck, Jr., "Persuasive Politics and Domination at Cahokia and Moundville," in *Leadership and Polity in Mississippian Society*, ed. Brian M. Butler and Paul D. Welch (Carbondale: Southern Illinois University Carbondale Center for Archaeological Investigations, 2006), 29; and James A. Brown, "The Cahokia Mound 72-Sub 1 Burials as Collective Representation," *Wisconsin Archeologist* 84, nos. 1–2 (2003): 83–99.

67 **"tools of political competition":** Jay Miller, *Ancestral Mounds: Vitality and Volatility of Native America* (Lincoln: University of Nebraska Press, 2015), 13.

4. Half Horse, Half Alligator

Michael Allen, *Western Rivermen, 1763–1861: Ohio and Mississippi Boatmen and the Myth of the Alligator Horse* (Baton Rouge: Louisiana State University Press, 1990), offers a thorough and highly readable portrait of the river's rough-and-tumble culture. Thomas Ruys Smith, *River of Dreams: Imagining in the Mississippi Before Mark Twain* effectively sets the river of the era in its national context. Walter Johnson, *River of Dark Dreams: Slavery and Empire in the Cotton Kingdom* (Cambridge, Mass.: Belknap Press of Harvard University Press, 2013), influenced my thinking on Thomas Jefferson and led me to Roger G. Kennedy, *Mr. Jefferson's Lost Cause: Land, Farmers, Slavery, and the Louisiana Purchase* (New York: Oxford University Press, 2003). The best and most readable biography of Black Hawk is Kerry R. Trask, *Black Hawk: The Battle for the Heart of America* (New York: Henry Holt, 2006).

71 **Prospect Robbins:** Richard L. Elgin, "The Initial Point of the 5th Principal Meridian," *American Surveyor,* April 2015; and Lynn Morrow, "A Surveyor's Challenges: P. K. Robbins in Missouri," Southeast Missouri State University Press, http://www.semopress.com/a-surveyors-challenges-p-k-robbins-in-missouri/.

72 **"inundations":** Quoted in Malcolm Rohrbough, *The Land Office Business: The Settlement and Administration of American Public Lands, 1789–1837* (New York: Oxford University Press, 1968), 81.

72 **"cypress and briers and thickets":** Quoted in the National Register of Historic Places for the site, filed with the U.S. Department of the Interior.

73 **"This noble and celebrated stream":** Zadok Cramer, *The Navigator* (Pittsburgh: Cramer & Spear, 1821), 122.

74 **"the distant roar of artillery":** Quoted in Louis C. Hunter, *Steamboats on the Western Rivers: An Economic and Technological History* (New York: Dover, 1993), chap. 5.

74 **sixty feet by fifteen:** Allen, *Western Rivermen,* 67

74 **"counterfeiters, horse thieves":** Cramer, *Navigator,* 247.

74 **"Wilson's Liquor Vault":** Paul Schneider, *Old Man River: The Mississippi in North American History* (New York: Picador, 2014), 228–29

74 **"empire for liberty":** Jefferson, in his letter to Clark, actually used the phrase "empire of liberty." In a later letter to James Madison, he reworked the phrase as "empire *for* liberty." See Johnson, *River of Dark Dreams,* 3–4, 24, 423n5.

75 **"cultivators":** Thomas Jefferson to John Jay, August 23, 1785, at https://avalon.law.yale.edu/18th_century/let32.asp.

75 **"husbandmen":** This phrase appears in Query 19 of Jefferson's *Notes on the State of Virginia,* though I've altered Jefferson's singular "husbandman" for the sake of grammar.

75 **"sores":** This reference also appears in *Notes on the State of Virginia.*

75 **"law of nature":** Thomas Jefferson, "Report on Negotiations with Spain," March 18, 1792, at https://founders.archives.gov/documents/Jefferson/01-23-02-0259.

76 **sixty feet long:** The statistics here are drawn from Allen, *Western Rivermen,* 66–69.

78 **"in turn extremely":** Timothy Flint, *Recollections of the Last Ten Years, Passed in Occasional Residences and Journeyings in the Valley of the Mississippi, from Pittsburg*

and the Missouri to the Gulf of Mexico, and from Florida to the Spanish Frontier (Boston: Cummings, Hilliard, 1826), 15.

78 **"dispute respecting":** Quoted in Smith, *River of Dreams*, 51.

79 **"helliferocious fellow":** Quoted in Lee Sandlin, *Wicked River: The Mississippi When It Last Ran Wild* (New York: Vintage, 2011), 99.

79 **"the first notable writer of fiction":** John T. Flanagan, "Morgan Neville, Early Western Chronicler," *Western Pennsylvania Historical Magazine*, December 1938, 255.

79 **"as did the Indian":** Quoted in John Banvard, "Description of Banvard's Panorama of the Mississippi River," at https://digital.lib.niu.edu/islandora/object/niu-twain%3A10956.

80 **"be glad to see":** Thomas Jefferson to William Henry Harrison, February 27, 1803, at https://founders.archives.gov/documents/Jefferson/01-39-02-0500.

80 **"above all others":** Quoted in Trask, *Black Hawk*, 29.

80 **"more like a civilized town":** Quoted in Trask, *Black Hawk*, 28

82 **"the poisonous effects":** *The Territorial Papers of the United States*, vol. 6: *The Territory of Mississippi, 1809–1817 Continued* (Washington, D.C.: U.S. Government Printing Office, 1938), 211.

82 **public sales could not be held:** Rohrbough, *Land Office Business*, 40

83 **"une grande carnage":** Quoted in Johnson, *River of Dark Dreams*, 21.

84 **"shut against":** Pekka Hämäläinen, *Lakota America* (New Haven, Conn.: Yale University Press, 2019), 152.

84 **"I have it much at heart":** Quoted in Smith, *River of Dreams*, 54.

84 **"the true spirit":** Quoted in Smith, *River of Dreams*, 45.

85 **"perhaps the most racially":** Smith, *River of Dreams*, 55.

85 **"ev'ry man":** Quoted in Smith, *River of Dreams*, 57.

5. The Office of River Improvement

Robert Gudmestad, *Steamboats and the Rise of the Cotton Kingdom* (Baton Rouge: Louisiana State University Press, 2011), and Gudmestad, "Steamboats and the Removal of the Red River Raft," *Louisiana History* 52, no. 4 (2011): 389–411, offer some of the more trustworthy details on Shreve's life.

88 **did more to change the ecosystem:** I borrow this claim from Gudmestad, *Steamboats and Rise*, 119.

88 **"pressing forward":** J.H.B. Latrobe, *The First Steamboat Voyage on the Western Waters* (Baltimore: Maryland Historical Society, 1871), at https://louisiana-anthology.org/texts/latrobe/latrobe--first_steamboat.html.

89 **"over blood heat":** Quoted in James L. Penick, *The New Madrid Earthquakes*, rev. ed. (Columbia: University of Missouri Press, 1981), 43.

89 **"One of the peculiar characteristics":** Latrobe, *First Steamboat Voyage*, n.p.

89 **white pine:** The description of the *New Orleans* is based on Gudmestad, *Steamboats and Rise*, 1–2.

90 **a 50 percent return:** Zadok Cramer, *The Navigator* (Pittsburgh: Cramer & Spear, 1821), 27.

91 **arrived in early 1815:** Decades later a biography would claim that Shreve arrived before the Battle of New Orleans and suggested he was key to the U.S. victory. Supposedly, he made a daring run downriver, through British artillery fire, to deliver munitions to a U.S. force. But there is solid evidence that Shreve arrived after the battle, and the biography in question is a sometimes-obvious work of hagiography. That this anecdote is now printed as fact in many encyclopedias and other reference books suggests how murky the historical Mississippi River remains. See Alfred A. Maass, "Brownsville's Steamboat Enterprize and Pittsburgh's Supply of General Jackson's Army," *Pittsburgh History* 77 (1994): 22–29.

91 **seventeen steamboats:** Margaret Jean Furrh, "Henry Miller Shreve: His Contribution to Navigation on the Western Rivers of the United States," master's thesis, Texas Tech University, 1971, 46.

92 **twenty days:** John O. Anfinson, *The River We Have Wrought: A History of the Upper Mississippi* (Minneapolis: University of Minnesota Press, 2003), 1–3.

93 **nearly fifty years:** Leland R. Johnson, *The Falls City Engineers: A History of the Louisville District* (U.S. Army Corps of Engineers, Louisville District, 1974), 46.

93 **"Uncle Sam's Toothpullers":** Todd Shallat, *Structures in the Stream: Water, Science, and the Rise of the U.S. Army Corps of Engineers* (Austin: University of Texas Press, 1994), 145.

94 **insurance rates for shipping dropped:** Gudmestad, *Steamboats and Rise*, 129.

94 **nearly 75,000 trees:** Louis C. Hunter, *Steamboats on the Western Rivers: An Economic and Technological History* (New York: Dover, 1993), chap. 4.

94 **twenty-four cords:** This and the other statistics about wood consumption and the disappearance of the American Bottom villages are from F. Terry Norris, "Where Did the Villages Go? Steamboats, Deforestation, and Archaeological Loss in the Mississippi Valley," in *Common Fields: An Environmental History of St. Louis*, ed. Andrew Hurley (St. Louis: Missouri History Museum, 1997), 73–89.

95 **thirty-seven miles of the bend:** James Barnett, *Beyond Control: The Mississippi River's New Channel to the Gulf of Mexico* (Jackson: University Press of Mississippi, 2017), 42.

95 **logjams that plugged 150 miles:** Gudmestad, "Steamboats and Removal," 391.

96 **thirty blankets and a few trinkets:** Gudmestad, "Steamboats and Removal," 403.

96 **earning $2.7 million:** Gudmestad, "Steamboats and Removal," 406.

97 **"soft" name:** See Claudio Saunt, *Unworthy Republic: The Dispossession of Native Americans and the Road to Indian Territory* (New York: W.W. Norton, 2020), xiii.

97 **23,000 Indigenous people:** Clyde Woods, *Development Arrested: The Blues and Plantation Power in the Mississippi Delta* (New York: Verso, 2017), 45.

97 **"The cold was unusually severe":** Alexis de Tocqueville, *Democracy in America* (1831), at https://www.gutenberg.org/cache/epub/815/pg815-images.html.

97 **fifteen thousand Choctaw:** Hannah Conway, "How Infrastructures Age: Engineering, Nature, and Environmental Justice in the Lower Mississippi Delta," Ph.D. diss., Harvard University (2023), 168.

97 **as much as $10 million:** Saunt, *Unworthy Republic*, 315.

98 **"the rich are the persons benefited":** Quoted in Johnson, *River of Dark Dreams,* 38.

98 **"Must have ice for their wine":** Frederick Law Olmsted, *The Cotton Kingdom: A Traveller's Observations on Cotton and Slavery in the American Slave States* (London: Sampson Low, Son & Co., 1862), vol. 2, at https://quod.lib.umich.edu/m/moa/AJA2492.0002.001?rgn=main;view=fulltext.

98 **Greenwood Leflore:** R. Halliburton, "Chief Greenwood Leflore and His Malmaison Plantation," in *After Removal: The Choctaw in Mississippi,* ed. Samuel J. Wells and Roseanna Tubby (Jackson: University Press of Mississippi, 1986), 56–63.

99 **revive the market:** Johnson, *River of Dark Dreams,* 32.

99 **common wisdom:** Mikko Saikku, *This Delta, This Land: An Environmental History of the Yazoo-Mississippi Floodplain* (Athens: University of Georgia Press, 2005), 104.

99 **Washington County:** The statistics about the enslaved population are from James C. Cobb, *The Most Southern Place on Earth: The Mississippi Delta and the Roots of Regional Identity* (New York: Oxford University Press, 1992), 8.

100 **"beautified the wilderness":** Quoted in Saikku, *This Delta,* 106.

100 **The rows were precise:** Johnson, *River of Dark Dreams,* 156–66.

100 **"thievish, poor, dirty":** Quoted in Kerry R. Trask, *Black Hawk: The Battle for the Heart of America* (New York: Henry Holt, 2006), 62.

101 **"fell like grass":** Quoted in Trask, *Black Hawk,* 7.

101 **"Our young men":** Quoted in Black Hawk, *Autobiography of Ma-ka-tai-me-she-kia-kiak, or Black Hawk,* Antoine LeClair, interpreter (1882), at https://digital.lib.niu.edu/islandora/object/niu-lincoln%3A35870.

102 **six million acres:** Roger L. Nichols, *Black Hawk and the Warrior's Path* (Malden, Mass.: John Wiley & Sons, 2017), 152.

102 **The boat spilled out workmen:** Mary Lethert Wingerd, *North Country: The Making of Minnesota* (Minneapolis: University of Minnesota Press, 2010), 135–36.

103 **"reclining lazily":** John Francis McDermott, "Banvard's Mississippi Panorama Pamphlets," *Papers of the Bibliographical Society of America* 43, no. 1 (1949): 48–62.

103 **"correct likenesses":** John Banvard, "Description of Banvard's Panorama of the Mississippi River," at https://digital.lib.niu.edu/islandora/object/niu-twain%3A10956..

103 **"a copious detail":** Quoted in Smith, *River of Dreams,* 105.

103 **"a sort of nightmare poem":** Johnson, *River of Dark Dreams,* 107.

103 **"gross and criminal mismanagement":** Quoted in Smith, *River of Dreams,* 105.

104 **"has contributed more":** Quoted in E. W. Gould, *Fifty Years on the Mississippi* (St. Louis: Nixon-Jones, 1889), 71.

104 **"brought within four days":** Quoted in Johnson, *River of Dark Dreams,* 78.

104 **half the country's citizens:** Gudmestad, *Steamboats and the Rise,* 5.

104 **more settlers than the famous gold rush:** Anfinson, *River We Have Wrought,* 5.

104 **the flatboats:** Michael Allen, *Western Rivermen, 1763–1861: Ohio and Mississippi Boatmen and the Myth of the Alligator Horse* (Baton Rouge: Louisiana State University Press, 1990), 63.

104 **howled back and forth:** Allen, *Western Rivermen,* 190.

104 **"There is something":** Quoted in Allen, *Western Rivermen,* 222.

105 **"This comes very *á propos*":** Quoted in Smith, *River of Dreams,* 131.

106 **Back in 1848:** Adam Kane, *The Western River Steamboat* (College Station: Texas A&M University Press, 2004), 26.

106 **Over a few years:** Johnson, *River of Dark Dreams,* 319.

107 **"Mr. Lincoln":** Quoted in Smith, *River of Dreams,* 182.

6. Dancing at the Skirts of Congress

John Barry's *Rising Tide: The Great Mississippi Flood of 1927 and How It Changed America* (New York: Simon & Schuster, 1997) is one of the great books on the river. It details Eads's and Humphreys's dispute so well that it leaves little new to say. Todd Shallat, *Structures in the Stream: Water, Science, and the Rise of the U.S. Army Corps of Engineers* (Austin: University of Texas Press, 1994), and Raymond H. Merritt, *Engineering in American Society: 1850–1875* (Lexington: University Press of Kentucky, 1969), offer helpful overviews of the Army Corps's position within U.S. engineering.

109 **"scarcely credible":** Quoted in Merritt, *Engineering in American Society,* 25.

110 **The label *engineer*:** Merritt, *Engineering in American Society,* 7.

110 **"a privileged order":** Quoted in Shallat, *Structures in the Stream,* 157.

110 **"some substantial good":** Quoted in Henry H. Humphreys, *Andrew Atkinson Humphreys: A Biography* (Philadelphia: John C. Winston Co., 1924), 35.

111 **"dancing attendance":** Quoted in Shallat, *Structures in the Stream,* 160.

111 **"very pleasant to deal with":** Quoted in Barry, *Rising Tide,* 34.

111 **Water lay atop:** Christopher Morris, *The Big Muddy: An Environmental History of the Mississippi and Its Peoples from Hernando de Soto to Hurricane Katrina* (New York: Oxford University Press, 2012), 106–7.

111 **"It is a work which":** Quoted in Humphreys, *Andrew Atkinson Humphreys,* 57.

112 **"excessive mental exertion":** Quoted in Barry, *Rising Tide,* 44.

112 **"The process by which":** Charles Ellet, Jr., *Report on Overflows of the Delta of the Mississippi* (Washington, D.C.: A. Boyd Hamilton, 1842), 43.

113 **"a delusive hope":** Ellet, *Report on Overflows,* 9.

113 **9.5 million acres:** George Pabis, "Delaying the Deluge: The Engineering Debate over Flood Control on the Lower Mississippi River, 1846–1861," *Journal of Southern History* 64, no. 3 (1998): 429.

113 **pushed a canoe:** Ann Vileisis, *Discovering the Unknown Landscape: A History of America's Wetlands* (Washington, D.C.: Island Press, 1997), 90.

113 **three to four feet tall:** Humphreys and Abbot, *Report Upon Physics,* 82–83.

114 **a fifth of the Louisiana sugar crop:** Vileisis, *Discovering the Unknown Landscape,* 80.

114 **"unwearied industry":** Quoted in Shallat, *Structures in the Stream,* 175.

114 **"enormous volume":** Humphreys and Abbot, *Report Upon the Physics,* 94.

114 **"every important fact":** Humphreys and Abbot, *Report Upon Physics,* 30.

115 **Within the decade:** Martin Reuss, "Andrew A. Humphreys and the Development

of Hydraulic Engineering: Politics and Technology in the Army Corps of Engineers, 1850–1950," *Technology and Culture* 26, no. 1 (1985): 1–33.

115 **"begun his survey":** Barry, *Rising Tide,* 53.

116 **"like a young girl":** Quoted in Barry, *Rising Tide,* 49.

116 **"anyone who knows me":** Humphreys, *Andrew Atkinson Humphreys,* 184.

116 **descendants of the early pioneers:** Macon Fry, *They Called Us River Rats: The Last Batture Settlement of New Orleans* (Jackson: University Press of Mississippi, 2021), 65.

118 **fixing up the rivers:** John O. Anfinson, *The River We Have Wrought: A History of the Upper Mississippi* (Minneapolis: University of Minnesota Press, 2003), 70–73.

118 **"If there were not":** Quoted in Edward Pross, "A History of Rivers and Harbors Appropriation Bills, 1866–1933," Ph.D. diss., Ohio State University (1938), 95.

118 **"In the rapid professionalization":** Merritt, *Engineering in American Society,* 4. Not all historians agree. John Davis, in "The U.S. Army Corps of Engineers and the Reconstruction of the American Landscape, 1865–1885," Ph.D. diss., Harvard University (2018), offers a compelling defense of the Corps's national role.

119 **mudlumps:** Mudlumps can stand as much as ten feet above the water surface. Some are small, not much bigger than a floating log; others can span acres. They appear to be caused by the weight of dense sands accumulating atop the underlying clay until it bursts upward, "rather like pastry dough in front of the chef's roller," as James M. Coleman et al. explain in "Sediment Instability in the Mississippi River Delta," *Journal of Coastal Research* 14, no. 3 (1998): 875. Some mudlumps were destroyed by waves within days of their appearance above the water's surface; others lasted for years. They're rarer now, thanks to the engineering on the river, but not gone entirely.

119 **"the most she can do":** Quoted in Barry, *Rising Tide,* 70.

121 **"until I could feel":** James Buchanan Eads, *Addresses and Papers* (St. Louis: Slawson & Co., 1884), 497.

121 **"entirely unsafe":** Quoted Barry, *Rising Tide,* 58. See also Merritt, *Engineering in American Society,* 125; John K. Brown, "Not the Eads Bridge: An Exploration of Counterfactual History of Technology," *Technology and Culture* 55, no. 3 (2014): 521–59.

121 **"phenomenon and eccentricity":** Quoted in Barry, *Rising Tide,* 77.

122 **"monster":** Quoted in Barry, *Rising Tide,* 65.

123 **Congress was happy enough:** They did make one major change: Eads had proposed building jetties at Southwest Pass, which was much wider (and deeper) and which Eads saw as the best route out of the river to accommodate the size of future oceangoing ships. Despite Eads's repeated entreaties, Congress commissioned the jetties at the shallower South Pass instead. Eads still believed he could deepen this pass, so he took the offer. Less than two decades later, as ships grew bigger, South Pass indeed proved too narrow. Today it's a quiet place. On a canoe trip to Eads's jetties, my companions and I encountered just a few shrimp boats and one man in a homemade boat who had motored down from Minnesota.

124 **"correct, permanently locate":** Quoted in Charles A. Camillo and Matthew T. Pearcy, *Upon Their Shoulders: A History of the Mississippi River Commission from Its Inception Through the Advent of the Modern Mississippi River and Tributaries Project* (Vicksburg: Mississippi River Commission, 2004), 37.

124 **his grand impossibilities:** In his sunset years, Eads's impossibilities grew increasingly wacky. When he died in 1887, he was pushing for a railroad line across a narrow Mexican isthmus that would take oceangoing freighters, lifted from the water, from the Gulf of Mexico to the Pacific.

124 **More money was spent:** Michael Allen, *Western Rivermen, 1763–1861: Ohio and Mississippi Boatmen and the Myth of the Alligator Horse* (Baton Rouge: Louisiana State University Press, 1990), 226.

124 **edge of Lake Winnibigoshish:** Calvin Fremling, *Immortal River: The Upper Mississippi in Ancient and Modern Times* (Madison: University of Wisconsin Press, 2005), 144–45. Some of the Ojibwe were hired on laborers.

125 **fourteen hours a day:** Leland R. Johnson, *The Davis Island Lock and Dam, 1870–1922* (U.S. Army Corps of Engineers, Pittsburgh District, 1985), 59. The accompanying dam was an innovative contraption, meant as a proof-of-concept: three hundred wooden "wickets," attached to the riverbottom with hinges, could lie flat when the river was high, allowing boats to use the full channel, rather than wait in line the lock. When the river dropped, a crew rode along in a boat, lifting the wickets one by one with grapples and cables, forming a wall that could hold back the water, creating a deep pool.

125 **fifty thousand people:** For an account of this day, see Johnson, *Davis Island Lock and Dam*, 94–97.

7. That Big Green Wall

Among the many books cited in this chapter, James C. Cobb's *The Most Southern Place on Earth: The Mississippi Delta and the Roots of Regional Identity* (New York: Oxford University Press, 1992), provides one of the most complete portraits of the plantation economy, though it's focused only on the Yazoo Basin. John C. Willis's *Forgotten Time: The Yazoo-Mississippi Delta After the Civil War* (Charlottesville: University Press of Virginia, 2000) provides another good look at the Yazoo Basin, in the late nineteenth century in particular, with especially important research into Black landowners.

127 **two stores:** This description of early Cleveland is drawn from Willis, *Forgotten Time*, 106.

127 **"Human life along the Mississippi":** Frederick Law Olmsted, *A Journey Through Texas: or, A Saddle-trip on the Southwestern Frontier* (New York: Dix, Edwards, 1857), 40.

128 **"that great Swamp":** A. A. Humphreys and H. L. Abbot, *Report Upon the Physics and Hydraulics of the Mississippi River* (Philadelphia: J.B. Lippincott, 1861), 24.

128 **more than doubled:** Mikko Saikku, *This Delta, This Land: An Environmental History of the Yazoo-Mississippi Floodplain* (Athens: University of Georgia Press, 2005), 107.

128 **"The river is looking":** Quoted in John Dean Davis, "Levees, Slavery, and Maintenance," *Technology's Stories* 6, no. 3 (2018): https://www.technologystories.org/levees-slavery-and-maintenance/.

128 **more than a hundred miles:** For the postwar destruction, see Ann Vileisis, *Discovering the Unknown Landscape: A History of America's Wetlands* (Washington, D.C.: Island Press, 1997), 81, and Cobb, *Most Southern Place on Earth*, 44.

129 **Within two years:** Will Dockery's story is one of many along this river that have become wrapped in legend. The best source I've found—the source most honest about the gaps in the record—is James W. Swinnich, *Living and Playing the Blues on Dockery Plantation-Farms* (n.p.: Outskirts Press, 2021).

129 **"The country was covered":** Quoted in Gaye Dean Wardlow et al., *King of the Delta Blues: The Life and Music of Charlie Patton* (Knoxville: University of Tennessee Press, 2022), 77.

129 **"chosen country":** Thomas Jefferson, "First Inaugural Address," March 4, 1801, at https://avalon.law.yale.edu/19th_century/jefinau1.asp.

129 **Nathaniel Bishop:** I'm indebted to the work of Thomas Ruys Smith, whose *Deep Water: The Mississippi River in the Age of Mark Twain* (Baton Rouge: Louisiana State University Press, 2019) provides a useful overview of the era's Mississippi River adventure writing.

130 **"the chains of every day of life":** Nathaniel Bishop, *Four Months in a Sneak-Box* (1879), at https://www.gutenberg.org/cache/epub/5686/pg5686.html.

130 **"so cramped up and sivilized":** Mark Twain, *The Adventures of Huckleberry Finn* (1884; reprint New York: Penguin, 2009), 35.

130 **"desolate farms and despondent individuals":** Quoted in Raymond H. Merritt, *Engineering in American Society: 1850–1875* (Lexington: University Press of Kentucky, 1969), 132.

131 **"terrible enemy":** Quoted in Charles A. Camillo and Matthew T. Pearcy, *Upon Their Shoulders: A History of the Mississippi River Commission from Its Inception Through the Advent of the Modern Mississippi River and Tributaries Project* (Vicksburg: Mississippi River Commission, 2004), 23.

131 **more than half of the Yazoo Basin:** John C. Hudson, "The Yazoo-Mississippi Delta as Plantation Country," in *Proceedings, Tall Timbers Ecology and Management Conference, February 22–24, 1979* (Thomasville, Ga.: Tall Timbers Research Station, 1982), 71.

131 **a few hundred acres:** See Willis, *Forgotten Time*, 63, 66.

131 **three-quarters of the landowners:** Cobb, *Most Southern Place on Earth*, 91.

132 **"We could hear wolves":** Mary Mann Hamilton, *Trials of the Earth: The True Story of a Pioneer Woman* (New York: Little, Brown, 2016), 81.

132 **90 percent of the land:** Willis, *Forgotten Time*, 105.

134 **"began with cannon fire":** This and the quote from Grant appear in Clyde Woods, *Development Arrested: The Blues and Plantation Power in the Mississippi Delta* (New York: Verso, 2017), 75.

135 **by the 1890s:** Camillo and Pearcy, *Upon Their Shoulders*, 90–92.

135 **"Over the years this great wall of earth":** Alan Lomax, *The Land Where the Blues Began* (New York: Pantheon, 1993), 212.

135 **"a few dollars":** Lomax, *Land Where the Blues Began*, 215.

136 **"wild work calls":** Lomax, *Land Where the Blues Began*, 212.

136 **One early Louisiana project:** Nathan Cardon, "'Less Than Mayhem': Louisiana's Convict Lease, 1865–1901," *Louisiana History* 58, no. 4 (2017): 434.

136 **"You could smell those camps":** Quoted in Lomax, *Land Where the Blues Began,* 234.

136 **"the last American frontier":** Lomax, *Land Where the Blues Began,* 216.

136 **"toughest places":** Quoted in Michael McCoyer, "'Rough Mens' in 'the Toughest Places I Ever Seen': The Construction and Ramifications of Black Masculine Identity in the Mississippi Delta's Levee Camps, 1900–1935," *International Labor and Working-Class History* 69, no. 1 (2006): 58.

136 **"that's all":** Quoted in Lomax, *Land Where the Blues Began,* 217.

136 **ordered Black prisoners:** Christine A. Klein and Sandra B. Zellmer, *Mississippi River Tragedies: A Century of Unnatural Disaster* (New York: NYU Press, 2014), 46.

137 **two to three times higher:** Richard M. Mizelle, "Black Levee Camp Workers, the NAACP, and the Mississippi Flood Control Project, 1927–1933," *Journal of African American History* 98, no. 4 (2013): 513, 521.

137 **"I can get all":** Quoted in McCoyer, "'Rough Mens,'" 76.

138 **"the river people":** This *Scribner's* piece is quoted in Gregg Andrews, *Shantyboats and Roustabouts: The River Poor of St. Louis, 1875–1930* (Baton Rouge: Louisiana State University Press, 2022), 17.

138 **At official meetings:** Lynn Morrow, "A 'Duck and Goose Shambles': Sportsmen and Market Hunters at Big Lake, Arkansas," Southeast Missouri State University Press, http://www.semopress.com/a-duck-and-goose-shambles-sportsmen-and-market-hunters-at-big-lake-arkansas/

139 **a knife to the ribs:** Roosevelt's hunt is described in Douglas Brinkley, *The Wilderness Warrior: Theodore Roosevelt and the Crusade for America* (New York: Harper Perennial, 2010), 432–40.

139 **"Beyond the end of cultivation":** Theodore Roosevelt, "In the Louisiana Canebrakes," *Scribner's Magazine* 43, no. 1 (1908): 47–60, at https://sites.rootsweb.com/~lamadiso/articles/louisianacanebrakes.htm.

139 **"A race of peaceful, unwarlike farmers":** Theodore Roosevelt, *Stories of the Great West* (New York: Century, 1916), 41.

140 **official Inland Waterways Commission inspection:** Mark Twain told the press he'd been asked to visit the tour but had declined; he did not approve of the engineers who were "annually sucking the blood of the Treasury" in their fruitless attempt to tame the river. Perhaps he was just venting his disdain for Roosevelt: the man was "still only fourteen years old after living a half century," Twain once wrote. He particularly despised the president's big-game hunting exploits. See Mark Twain, *Autobiography of Mark Twain* (Oakland: University of California Press, 2015), 3:136, 175.

140 **dropped by half:** Leland R. Johnson, *The Falls City Engineers: A History of the Louisville District* (U.S. Army Corps of Engineers, Louisville District, 1974), 178.

141 **so little traffic:** John Ferrell, *Soundings: 100 Years of the Missouri River Navigation Project* (U.S. Army Corps of Engineers, 1996), 44.

141 **no through traffic:** John O. Anfinson, *The River We Have Wrought: A History of the Upper Mississippi* (Minneapolis: University of Minnesota Press, 2003), 188.

141 **"psychopathic enthusiasts":** Quoted in Karen O'Neill, *Rivers by Design: State Power and the Origins of U.S. Flood Control* (Durham, N.C.: Duke University Press, 2006), 137.

141 **Harold Speakman:** The book is *Mostly Mississippi* (New York: Dodd, Mead, 1927). For another view of the river in this era, see Kent Lighty and Margert Lighty, *Shantyboat* (New York: Century, 1930).

141 **"You could hear":** Wallace Stegner and Pete Stegner, *American Places* (Moscow: University of Idaho Press, 1983), 67.

142 **This special rule:** Martin Reuss, *Designing the Bayous: The Control of Water in the Atchafalaya Basin, 1800–1995* (College Station: Texas A&M University Press, 2004), 94.

142 **"We levee contractors":** Quoted in Lomax, *Land Where the Blues Began*, 229.

142 **a new province:** More than cotton grew along the river, though. Sugar, after slumping briefly in the postwar years, had returned to the delta in Louisiana. In Arkansas, meanwhile, in the prairies along the western edge of the floodplain, there were 180,000 acres of rice in 1920. See Pete Daniels, *Breaking the Land: The Transformation of Cotton, Tobacco, and Rice Cultures Since 1880* (Champaign: University of Illinois Press, 1986), 58.

142 **"probably more firmly fixed":** Quoted in Cobb, *Most Southern Place on Earth*, 98.

143 **Some charged tenants:** Cobb, *Most Southern Place on Earth*, 101–2.

143 **incorporated in late 1907:** John C. Fisher, *Southeast Missouri from Swampland to Farmland: The Transformation of the Lowlands* (Jefferson, N.C.: McFarland & Co., 2017), 113, 120, 127–29.

143 **"The road to wealth":** Southern Alluvial Land Association, *The Call of the Alluvial Empire* (Memphis: Southern Alluvial Land Association, 1920), 7.

144 **"all but useless":** Quoted in Hannah Conway, "How Infrastructures Age: Engineering, Nature, and Environmental Justice in the Lower Mississippi Delta," Ph.D. diss., Harvard University (2023), 74.

145 **"among the bull frogs":** Quoted in Swinnich, *Living and Playing*, 114.

145 **"so fast now":** William Faulkner, *Big Woods: The Hunting Stories* (New York: Vintage Books, 1994), 165.

8. The Great Flood

Useful accounts of the 1927 flood include John Barry, *Rising Tide: The Great Mississippi Flood of 1927 and How It Changed America* (New York: Simon & Schuster, 1997), and Pete Daniel, *Deep'n as It Come: The 1927 Mississippi River Flood* (Fayetteville: University of Arkansas Press, 1996). The Jadwin plan and its aftermath are explored in Charles A. Camillo and Matthew T. Pearcy, *Upon Their Shoulders: A History of the Mississippi River Commission from Its Inception Through the Advent of the Modern Mississippi River and Tributaries Project* (Vicksburg: Mississippi River Commission, 2004), 141–205, and in Martin Reuss, *Designing the Bayous: The Control of Water in the Atchafalaya Basin, 1800–1995* (College Station: Texas A&M University Press, 2004), 103–34. Various Army Corps reports provide the details of the Mississippi Basin Model, especially J. E. Foster, "History and Description of the Mississippi Basin Model," *Mississippi Basin Model Report 1-6* (Vicksburg: U.S. Army Engineer Waterways Experiment Station, 1971), and Ben H. Fatherree, *The First 75 Years: History of Hydraulics Engineering and the Waterways Experiment Station* (Vicksburg: U.S. Army Engineer Research and Development Center, 2004).

149 **In March:** The statistics and figures in these paragraphs are drawn from Barry, *Rising Tide,* 182, 187, 196.

150 **a small crevasse:** A list of crevasses can be found in H. C. Frankenfield, "The Floods of 1927 in the Mississippi Basin," *Monthly Weather Review,* Supplement no. 29, U.S Department of Agriculture Weather Bureau (Washington, D.C.: U.S. Government Printing Office, 1927), 34.

150 **More than 22 million gallons:** Unless otherwise noted, details about the Mounds Landing crevasse in the following pages are drawn from Barry, *Rising Tide,* 198–203, 206, 275, 278–79.

150 **"We can't hold it":** Quoted in Barry, *Rising Tide,* 201.

151 **"overflow":** Quoted in Barry, *Rising Tide,* 201.

151 **The sound of the water:** Daniel, *Deep'n as It Come,* 6.

151 **"a glittering, slimy mass":** Lucy Somerville, "The Mississippi Flood of 1927," Mississippi History Now, https://www.mshistorynow.mdah.ms.gov/issue/the-mississippi-flood-of-1927.

152 **"the corn-liquor center":** Quoted in Kerry Rose Dicks, "Against Current: On River Bootleggers, Bayou Moonshiners, and the Perseverance of Prohibition in Mississippi—Part II," *Bluffs and Bayous* (2018): 51.

152 **"pitched battle":** Quoted in Barry, *Rising Tide,* 192.

152 **"a highway of humanity":** Quoted in Andy Horowitz, *Katrina: A History, 1915–2015* (Cambridge, Mass.: Harvard University Press, 2020), 23.

153 **more than two hundred:** Frankenfield, "The Floods of 1927," 34.

153 **Water covered 16.5 million:** Daniel, *Deep'n as It Come,* 8.

153 **Nearly six hundred thousand residents:** Daniel, *Deep'n as It Come,* 120–21.

153 **more than half:** Daniel, *Deep'n as It Come,* 120–21.

153 **Estimates vary:** Charles A. Camillo, *Divine Providence: The 2011 Flood in the Mississippi River and Tributaries Project* (Vicksburg: Mississippi River Commission, 2012), 16.

153 **fifteen cents was spent on daily rations:** Barry, *Rising Tide,* 328.

153 **"teach them":** Quoted in Barry, *Rising Tide,* 328.

154 **"a plaintive dirge":** T. H. Alexander, "Hoover Now Hero of Flood South," *New York Times,* July 31, 1927.

155 **"You do not expect":** Quoted in Barry, *Rising Tide,* 404.

155 **"That is news":** Quoted in Barry, *Rising Tide,* 405.

155 **"the greatest engineering feat":** Quoted in Barry, *Rising Tide,* 406.

156 **"back to the mines":** Quoted in Jill Lepore, *These Truths: A History of the United States* (New York: W.W. Norton, 2018), 424.

158 **"Lazy Man's Reach":** Albert Tousley, *Where the River Goes* (Iowa City: Tepee Press, 1928), 210.

158 **"Four 'drifts' downstream":** Tousley, *Where the River Goes,* 211.

159 **"pipeline":** Quoted in Camillo and Pearcy, *Upon Their Shoulders,* 192.

159 **"The best that the laboratory can do":** Quoted in Camillo and Pearcy, *Upon Their Shoulders,* 202.

160 **"Do you want me to":** Camillo, *Divine Providence,* 148.

160 **"Nobody even noticed much":** Quoted in Camillo, *Divine Providence,* 150.

160 **the largest river channelization program:** Paul Hudson, *Flooding and Management of Large Fluvial Lowlands: A Global Environmental Perspective* (New York: Cambridge University Press, 2021), 156.

160 **One report:** Ernest Theodore Hiller, *Houseboat and River Bottoms People: A Study of 684 Households in Sample Localities Adjacent to the Ohio and Mississippi Rivers* (Urbana: University of Illinois Press, 1939), 14.

161 **"proper activity":** Quoted in Joseph L. Arnold, *The Evolution of the 1936 Flood Control Act* (Fort Belvoir, Va.: Office of History, U.S. Army Corps of Engineers, 1988), 70.

163 **"among the largest":** Quoted in Boyce Upholt, "The Mississippi River Is Under Control—For Now," *Time*, August 7, 2019.

165 **a hypothetical apocalypse:** Camillo, *Divine Providence*, 6–7.

166 **nearly two thousand captive soldiers:** Foster, "History and Description of Basin Model," 5.

168 **$65 million:** Fatherree, *The First 75 Years*, 93.

9. Hooked Up, Hard Down

The ecological impacts of dam construction along the Upper Mississippi are well documented in John O. Anfinson, *The River We Have Wrought: A History of the Upper Mississippi* (Minneapolis: University of Minnesota Press, 2003), and Calvin Fremling, *Immortal River: The Upper Mississippi in Ancient and Modern Times* (Madison: University of Wisconsin Press, 2005). Reports from the U.S. Geological Survey bring the picture up to date: see Jeffrey H. Houser, *Ecological Status and Trends of the Upper Mississippi and Illinois Rivers* (Reston, Va.: U.S. Geological Survey, 2022) and Nathan R. de Jager et al., *Indicators of Ecosystem Structure and Function for the Upper Mississippi River System* (Reston, Va.: U.S. Geological Survey, 2018). The still-wild status of the Lower Mississippi, meanwhile, is explored in D. S. Biedenharn et al., *Attributes of the Lower Mississippi River Batture,* Mississippi River Geomorphology and Potamology Program Technical Note No. 4 (U.S. Army Corps of Engineers, 2018), and in Paul Hartfield, "Engineered Wildness: The Lower Mississippi River, an Underappreciated National Treasure," *International Journal of Wilderness* 20, no. 3 (2014): 8–13.

173 **"a primary center of diversity":** Henry W. Robison, "Zoogeographic Implications of the Mississippi River Basin," in *The Zoogeography of North American Freshwater Fish*, ed. C. H. Hocutt and E. O. Wiley (New York: John Wiley & Sons), 27.

174 **nine feet deep:** South of Baton Rouge the channel is maintained to a depth of 50 feet, to accommodate deeper-draft oceangoing freighters.

174 **As many as 150 square miles of riverine habitat:** Brandon J. Sansom et al., "Performance Evaluation of a Channel Rehabilitation Project on the Lower Missouri River and Implications for the Dispersal of Larval Pallid Sturgeon," *Ecological Engineering* 194 (2023): 194.

174 **gone from their ancient home:** See William R. Ardren et al., "Demographic and Evolutionary History of Pallid and Shovelnose Sturgeon in the Upper Missouri

River," *Journal of Fish and Wildlife Management* 13, no. 1 (2022): 124–43. Jacobson indicated that while many hatchery-raised pallid sturgeon live in the Upper Missouri, the lack of wild fish may mean a problematic decline in genetic diversity.

175 **Now it was blocked:** The language-blocking interception-rearing complexes appeared in the 2020 Water Resources Development Act (WRDA). Jacobson, who retired from the U.S. Geological Survey in 2023, indicated that a draft version of the required study confirmed that any impacts on navigation were within the range of error of the models used. The 2022 WRDA authorized the construction of two more IRCs, but given the opposition to IRCs, the Corps's Missouri River Recovery Program is seeking other options. This may open the Corps to new litigation over its failure to protect the pallid sturgeon, as the IRC program was a required component of its compliance with the Endangered Species Act.

175 **Within a decade:** Leland R. Johnson, *The Falls City Engineers: A History of the Louisville District* (U.S. Army Corps of Engineers, Louisville District, 1974), 189.

176 **"modernized":** Johnson, *Falls City Engineers,* 233

176 **local traffic increased tenfold:** Anfinson, *River We Have Wrought,* 276.

176 **"towboatingist":** George Horne, "Leading Towboat City Gives One Man All Credit," *New York Times,* July 26, 1964.

178 **literal descendants of shantyboaters:** Melody Golding, *Life Between the Levees: America's Riverboat Pilots* (Jackson: University Press of Mississippi, 2019), offers useful oral histories with towboat pilots, including this detail.

179 **U.S. inland rivers:** Waterways Council offers these statistics at https://waterwayscouncil.org/waterways-system.

179 **traffic hit a peak:** National Research Council, *The Missouri River Ecosystem: Exploring the Prospects for Recovery* (Washington, D.C.: National Academies Press, 2002), 6.

179 **most subsidized form:** Congressional Budget Office, *Financing Waterways Development: The User Charge Debate* (Washington, D.C.: Congressional Budget Office, 1977).

180 **even human feces:** In 1900 the city of Chicago opened a canal that linked Lake Michigan to the Illinois River, meant to steer its sewage south toward the Mississippi watershed, rather than north into the lake, the source of its drinking water. The Illinois joins the Mississippi just above St. Louis, where locals were not particularly pleased to be receiving the waste of the country's then-second-largest city. The canal became the subject of the first water-pollution case considered by the U.S. Supreme Court, which ruled in favor of Chicago in 1906. The judges decided that the canal carried so much clean water out of Lake Michigan that the Illinois, once "sluggish and ill smelling," had actually been turned into a clear stream. Besides, they noted, the city of St. Louis dumped its sewage into the Mississippi River, too, so Missouri wasn't in much of a position to complain. Quoted in Dan Egan, *The Death and Life of the Great Lakes* (New York: W.W. Norton, 2017), 164.

181 **failed to pass any major law:** National Research Council, *New Directions in Water Resources Planning for the U.S. Army Corps of Engineers* (Washington, D.C.: National Academies Press, 2009), 17.

181 **"a nationally significant":** Quoted in Anfinson, *River We Have Wrought,* 288.

181 **described them as mountains:** Fremling, *Immortal River*, 79.

182 **By some reports:** John Madson, *Up on the River* (New York: Schocken Books, 1985), 264.

184 **240 square miles gone:** See De Jager et al., *Ecological Status and Trends*, 6, 16.

185 **more and longer flooding:** Madeline Heim, "High Water and Prolonged Flooding Are Changing the Ecosystem of the Upper Mississippi River, a New Report Finds," *Milwaukee Journal Sentinel*, June 27, 2022.

185 **hundreds of millions of dollars:** Even this wound up being a substantial underestimate. When the lock was finally constructed, costs ballooned from the budgeted $350 million to more than a billion. See National Research Council, *Inland Navigation System Planning: The Upper Mississippi-Illinois Waterway* (Washington, D.C.: National Academies Press, 2001), 48.

185 **90 percent or more of the system's costs:** Tyler J. Kelley, *Holding Back the River: The Struggle Against Nature on America's Waterways* (New York: Avid Reader Press, 2021), 49.

185 **The system required:** Jay Reeves, "$2B Waterway Through Deep South Yet to Yield Promised Boom," Associated Press, September 16, 2019.

185 **The installation:** "Factors Contributing to Cost Increases and Schedule Delays in the Olmsted Locks and Dam Project," Government Accountability Office, February 2017.

186 **to solve the problem:** Charles V. Stern et al., "Water Resource Issues in the 117th Congress," Congressional Research Service, March 12, 2021.

186 **$7 billion to $9 billion:** Agribusiness Consulting, "Importance of Inland Waterways to U.S. Agriculture," prepared for the U.S. Department of Agriculture Agricultural Marketing Department, August 2019, 1.

186 **coal and petroleum:** Agribusiness Consulting, "Importance of Inland Waterways," 17. The figures are from 2013–17. If you look at ton-miles, instead of just short tons, fossil fuels drop to a third of the cargo, indicating that these goods are not carried as far along the rivers as grain and other farm products.

186 **"well-being of the Midwest":** Quoted in Michael Grunwald, "How Corps Turned Doubt Into a Lock," *Washington Post*, February 13, 2000.

187 **"Program Growth Initiative":** For Sweeney's tale, including the orders quoted here, see Grunwald, "How Corps Turned Doubt." For Program Growth Initiative, see Grunwald, "Generals Push Huge Growth for Engineers," *Washington Post*, February 24, 2000.

187 **one of Grunwald's editors joked:** Michael Grunwald, "Journalist Michael Grunwald on the Hubris of the Army Corps," *Grist*, March 19, 2008.

187 **"serious misconduct":** Quoted in Nicole Carter, *Army Corps of Engineers: Civil Works Reform Issues for the 107th Congress*, Congressional Research Service, March 27, 2002.

187 **a range of future scenarios:** Grunwald, "Journalist Michael Grunwald."

187 **"outlandishly optimistic":** Grunwald, "Journalist Michael Grunwald."

188 **"green pork":** Grunwald, "Journalist Michael Grunwald."

188 **it's slowly being built:** When the Government Accountability Office reviewed the program in 2020, it found that $25.2 million had been allocated to ecosystem proj-

ects, compared to $32.1 million for navigation. See "U.S. Army Corps of Engineers: Information on the Navigation and Ecosystem Sustainability Program," Government Accountability Office, January 22, 2021, https://tinyurl.com/5n7xh6xu. Most of that work was just on developing designs. The first full project of any kind was completed in 2023: a notch in a wing dam on a lock in Minnesota that is meant to produce more fish habitat. The advanced age of the locks continues to impact the navigation industry; according to one 2019 study, half the boats on the Mississippi River get hung up at locks in delays that can last as long as two hours. The Corps of Engineers has calculated that at just two locks, such slowdowns cost up to $35 million per year.

188 **"fun fact!":** *2016 Report to Congress: Upper Mississippi River Restoration Program* (U.S. Army Corps of Engineers, Rock Island District, 2016), 52.

188 **Asian carp now make up nearly two-thirds:** De Jager et al., *Ecological Status and Trends*, 183, 194.

189 **Portions of Pool No. 8:** "Minnesota, Wisconsin Conduct Carp Harvest," *Waterways Journal*, April 21, 2021.

190 **They've emphasized Congress's:** Lynn Muench et al. to Elliott Stefanik, January 23, 2015, at https://tinyurl.com/shwbtv59. The quote on tonnage appears in U.S. Congress, House Committee on Transportation and Infrastructure, *Water Resources Reform and Development Act of 2013: Report of the Committee on Transportation and Infrastructure to Accompany H.R. 3080 Together with Additional Views, Part 1* (Washington, D.C.: U.S. Government Printing Office, 2013), 66.

190 **tossed up their hands:** Kat McCain, Sara Schmuecker, and Nathan R. De Jager, *Habitat Needs Assessment-II for the Upper Mississippi River Restoration Program: Linking Science to Management Perspectives* (U.S. Army Corps of Engineers, Rock Island District, 2018), 12.

191 **began to cut notches:** The dike-notching strategy had been tried and abandoned before. First authorized by the same 1986 Water Resources Development Act that launched restoration on the Upper Mississippi, the program was abandoned in the 1990s over worries that the size of notches that fish required might cause a "detrimental realignment" of the navigation channel. See Robert Kelly Schneiders, *Unruly River: Two Centuries of Change Along the Missouri* (Lawrence: University Press of Kansas, 1999), 246.

191 **by causing more flooding:** The science is complicated. The stretch of river in question, between Omaha, Nebraska, and St. Joseph, Missouri, has been known as a "low conveyence zone"—a reach that struggles to successfully carry high discharges within its banks—since before the habitat restoration program was launched. Before the Missouri was channelized, the Corps of Engineers suggested that the levees be set as much as five thousand feet apart, creating a local floodway. But farmers often cleared land nearly all the way to the riverbanks, increasing the threat of floods. The farmers will be awarded onetime payments for flowage easements, so the government will not need to pay again for future floods. See *Ideker Farms, Inc. v. United States*, No. 21-1849 (Fed. Cir. 2023). Litigation remains ongoing; the federal government could appeal and seek a Supreme Court ruling, and the prosecuting attorneys have already filed a class-action suit they describe as a "sequel."

192 **"delisted" across its range:** These days, though, the birds seem to be traveling up

the Ohio River rather than the Missouri—a signal, perhaps, of their opinion of the local river "improvements" on the troubled Missouri.

192 **eighty-four backchannels:** U.S. Fish and Wildlife Service, *Biological Opinion: Channel Improvement Program, Mississippi River and Tributaries Project, Lower Mississippi River* (Jackson, Miss.: U.S. Fish and Wildlife Service, Mississippi Field Office, 2013), 55.

193 **Nearly two hundred such islands:** Paul Hudson et al., "(Re)Development of Fluvial Islands Along the Lower Mississippi River over Five Decades, 1965–2015," *Geomorphology* 331 (2019): 78–91.

193 **Clean Water Act:** See R. Eugene Turner, "Declining Bacteria, Lead, and Sulphate, and Rising pH and Oxygen in the Lower Mississippi River," *Ambio* 50 (2021): 1731–38.

194 **among the highest concentrations of plastics:** Tristan Baurick, "Gulf of Mexico Loaded With One of World's Highest Concentrations of Plastic; Here's Why," *New Orleans Times-Picayune*, August 16, 2017.

194 **"one of the most important remaining wilderness areas":** Biedenharn et al., *Attributes of Lower Mississippi River Batture*, 2.

10. Death Alley

Historian Pete Daniel has written several excellent books documenting the social and environmental consequences of twentieth-century agriculture in the alluvial valley. *Dispossession: Discrimination Against African American Farmers in the Age of Civil Rights* (Chapel Hill: University of North Carolina Press, 2013) examines how New Deal programs were reconfigured as weapons of white supremacy, and *Toxic Drift: Pesticides and Health in the Post-World War II South* (Baton Rouge: Louisiana State University Press, 2005) explores the rise of chemical agriculture.

197 **"nothing but flatness":** Quoted in Peggy Whitman Prenshaw, ed., *Conversations with Eudora Welty* (Jackson: University Press of Mississippi, 1984), 82.

200 **golden plovers:** David Muth, "The Once and Future Delta," in *Perspectives on the Restoration of the Mississippi Delta*, ed. John W. Day et al. (Dordrecht, Netherlands: Springer, 2014), 11.

201 **Consider the Little River:** For a good historical overview of the district, see John C. Fisher, *Southeast Missouri from Swampland to Farmland: The Transformation of the Lowlands* (Jefferson, N.C.: McFarland & Co., 2017).

202 **more than half the landowners:** These struggles appear in Fisher, *Southeast Missouri*, 136–50.

202 **tens of thousands of tenant farmers:** Alison Collis Greene, *No Depression in Heaven: The Great Depression, the New Deal, and the Transformation of Religion in the Delta* (New York: Oxford University Press, 2016), 11.

202 **$1.5 billion:** Pete Daniel, *Breaking the Land: The Transformation of Cotton, Tobacco, and Rice Cultures Since 1880* (Champaign: University of Illinois Press, 1986), 92.

202 **the planters here swallowed:** Nationwide, in 1933, of the payments in excess of $10,000, 25 percent went to the Yazoo Basin. Over the next two years, that figure rose

to 35 percent and then 41 percent. James C. Cobb, *The Most Southern Place on Earth: The Mississippi Delta and the Roots of Regional Identity* (New York: Oxford University Press, 1992), 192.

202 **$167,000:** Jeannie M. Whayne, *Delta Empire: Lee Wilson and the Transformation of Agriculture in the New South* (Baton Rouge: Louisiana State University Press, 2011), 180–81.

202 **"People come in here":** Quoted in Cobb, *Most Southern Place on Earth*, 195–96.

203 **the Citizens' Council:** Nathan A. Rosenberg and Bryce Wilson Stucki, "The Butz Stops Here: Why the Food Movement Needs to Rethink Agricultural History," *Journal of Food Law and Policy 13, no.* 1 (2018): 16.

203 **"visible evidence":** Cobb, *Most Southern Place on Earth*, 256.

203 **fell by 78 percent:** Christopher Morris, *The Big Muddy: An Environmental History of the Mississippi and Its Peoples from Hernando de Soto to Hurricane Katrina* (New York: Oxford University Press, 2012), 181.

203 **tacked onto existing farms:** Robert W. Harrison, *Alluvial Empire* (Little Rock: Delta Fund and U.S. Department of Agriculture, 1961), 1:309.

204 **"Since the earth":** Quoted in Greene, *No Depression in Heaven*, 178.

204 **"un-American social experiments":** Quoted in Donald Holley, *Uncle Sam's Farmers: The New Deal Communities in the Lower Mississippi Valley* (Chicago: University of Illinois Press, 1975), 248.

205 **in a typical year:** These figures appear in Cobb, *Most Southern Place on Earth*, 259.

205 **"giving people something for nothing":** Quoted in Cobb, *Most Southern Place on Earth*, 261.

205 **A research team:** Van Newkirk II, "The Great Land Robbery," *Atlantic*, August 2019.

205 **"new type of black ghetto":** Charles S. Aiken, "A New Type of Black Ghetto in the Plantation South," *Annals of the Association of American Geographers* 80, no. 2 (1990): 223.

205 **hedge funds and other institutional investors:** Tom Philpott, *Perilous Bounty: The Looming Collapse of American Farming and How We Can Prevent It* (New York: Bloomsbury, 2020), 61–62; Newkirk, "Great Land Robbery."

206 **$1.5 billion in crop insurance:** Anne Schechinger, "In the Mississippi River Region," Environmental Working Group, March 23, 2022, https://www.ewg.org/research/mississippi-river-region-billions-dollars-spent-crop-insurance-payouts-could-have-been.

206 **little of the value:** Jeremy G. Weber et al., "Crop Prices, Agricultural Revenues, and the Rural Economy," *Applied Economic Perspectives and Policy* 37, no. 3 (2015): 459–76.

206 **in much of the alluvial valley:** Raj Chetty et al., "Where Is the Land of Opportunity? The Geography of International Mobility in the United States," Working Paper no. 19843, National Bureau of Economic Research (2014), 3. "Some [commuting zones] in the U.S. . . . have lower levels of mobility than any developed country for which data are available," the authors write. Raj Chetty et al., "The Opportunity Atlas: Mapping the Childhood Roots of Social Mobility," https://opportunityinsights.org/paper/the-opportunity-atlas/, offers maps that show which counties have the lowest levels of mobility.

206 **Shawn Peebles:** For more of Peebles's story, see Travis Lux, "Hot Farm Part 4: The New California," Food and Environmental Reporting Network, 2022. If other farmers switch to vegetable farming en masse—something the World Wildlife Fund is encouraging, in an effort to avoid new deforestation elsewhere as vegetable production shifts away from climate-ravaged California—the change will have implications for river navigation, too. Unlike grain, vegetables need cold storage and would be unlikely to be shipped by barge.

207 **Roughly 10 percent:** Thomas Turnbull et al., "Quantifying Available Energy and Anthropogenic Energy Use in the Mississippi River Basin," *Anthropocene Review* 8, no. 3 (2021): 294.

207 **By 1937:** Jack E. Davis, *The Gulf: The Making of an American Sea* (New York: Liveright, 2017), 301.

208 **Just this narrow sliver of river:** Halle Parker, "This Sliver of Louisiana Is Responsible for Most of the State's Carbon Emissions, New Study Says," *WWNO*, July 6, 2022.

208 **spits out a second ton:** Kristen Mosbrucker, "St. James Parish Alumina Smelter Investor Buys Much Bigger Stake; Gobbling Up Jamaican Mine," *Advocate*, July 14, 2021.

208 **a pile of phosphogypsum powder:** Boyce Upholt, "Eaters of the Earth: How the Fertilizer Industry Leaves a Trail of Destruction Across the American South," *Counter*, April 20, 2020.

209 **"the application of human energy":** Walter Johnson, *River of Dark Dreams: Slavery and Empire in the Cotton Kingdom* (Cambridge, Mass.: Belknap Press of Harvard University Press, 2013), 164.

210 **leaky refinery in Baton Rouge:** These Louisiana disasters are chronicled in Raymond J. Burby, "Baton Rouge: The Making (and Breaking) of a Petrochemical Paradise," in *Transforming New Orleans and Its Environs*, ed. Craig Colten (Pittsburgh: University of Pittsburgh Press, 2000), 160–77.

210 **Bayou Corne:** Katy Reckdahl, "When the Ground Gives Way," *Places*, August 2019.

211 **as much as $200,000:** Keith Schneider, "Chemical Plaznts Buy Up Neighbors for Safety Zone," *New York Times*, November 28, 1990.

212 **conducted a survey:** Michael Orr, interview by the author, June 1, 2021.

212 **"bad sunburn":** Quoted in Mara Kardas-Nelson, "The Petrochemical Industry Is Killing Another Black Community in 'Cancer Alley,'" *Nation*, August 26, 2019.

213 **During the 1957 spraying season:** Daniel, *Toxic Drift*, 19–21.

213 **as much as threefold:** Nancy N. Rabalais et al., "Characterization of Hypoxia: Topic 1 Report for the Integrated Assessment on Hypoxia in the Gulf of Mexico," NOAA Coastal Ocean Program Decision Analysis Series no. 15 (1999).

214 **paid out billions in settlements:** The company "stands fully behind" the safety of its products, as it declares on a statement on its website. Briefly, the legal tide seemed to have turned in Bayer's favor: the company won seven consecutive jury trials focused on the glyphosate, until losing several cases in late 2023.

214 **In the 2017 growing season alone:** Boyce Upholt, "A Killing Season," *New Republic*, December 2018.

214 **trained volunteers from Arkansas Audubon:** Andy McGlashen, "Bitter Harvest," *Audubon,* Winter 2022.

215 **"It's starting to get":** Quoted in Emily Unglesbee, "Dicamba-Resistant Pigweed," *Progressive Farmer/DTN,* July 28, 2020. See also Unglesbee, "Beware Zombie Weeds," *Progressive Farmer/DTN,* July 3, 2021.

216 **running low more often:** Lindsey M. W. Yasarer et al., "Trends in Land Use, Irrigation, and Streamflow Alteration in the Mississippi River Alluvial Plain," *Frontiers in Environmental Science* 8 (2020), https://www.frontiersin.org/articles/10.3389/fenvs.2020.00066/full.

216 **Researchers are scrambling:** Boyce Upholt, "An Interstate Battle for Groundwater," *Atlantic,* December 4, 2015.

216 **aquifers in the Yazoo Basin:** Meredith L. Brock et al., "Evaluation of On-Farm Water Capture and Groundwater Decline in the Big Sunflower Watershed, Mississippi River Basin," *Journal of Hydrology: Regional Studies* 48 (2023): 13.

11. The Great Unraveling

The classic—and painfully well-written—account of the trouble at Old River is John McPhee's essay "Atchafalaya," which appears in McPhee, *The Control of Nature* (New York: Farrar, Straus & Giroux, 1989). Other useful resources include James Barnett, *Beyond Control: The Mississippi River's New Channel to the Gulf of Mexico* (Jackson: University Press of Mississippi, 2017); Martin Reuss, *Designing the Bayous: The Control of Water in the Atchafalaya Basin, 1800–1995* (College Station: Texas A&M University Press, 2004); and James W. Lewis and Jonathan A. Ashley, *Technical Assessment of the Old, Mississippi, Atchafalaya, and Red (OMAR) Rivers: Main Report,* Mississippi River Geomorphology and Potamology Technical Program Report no. 41, vol. 1 (U.S. Army Corps of Engineers, 2022).

219 **gathered every morning in a New Orleans conference room:** Boyce Upholt, "The Mississippi River Is Under Control—For Now," *Time,* August 7, 2019.

220 **Economists calculated:** Upholt, "River Is Under Control—For Now." In 2023, a judge ruled the army must consult with federal fisheries officials before opening the gates.

220 **lately the river's behavior had changed:** Due to several issues that I discuss in this chapter, the same discharge now often produces higher water surface elevations on the Mississippi River. This effect was exacerbated in 2019 by the fact that a flood crest pouring down an already flooded river leads to further increases. Within a few days of the spillway opening, the river did reach the official trigger of 1.25 million cubic feet per second.

221 **fruit ranch:** Richard J. Russell, "Memorial to Harold Norman Fisk (1908–1964)," *GSA Bulletin* 76, no. 4 (1965): 53.

221 **"reacted violently":** Rufus J. Leblanc, "Harold Normal Fisk as a Consultant to the Mississippi River Commission, 1948–1964—An Eye-Witness Account," *Engineering Geology* 45 (1996): 24.

223 **"poised stream":** Harold Fisk, *Geological Investigation of the Alluvial Valley of the Lower Mississippi River* (Vicksburg: Mississippi River Commission, 1944), 50, 56.

224 **"Until an engineering structure":** Paul Hudson, *Flooding and Management of Large Fluvial Lowlands: A Global Environmental Perspective* (New York: Cambridge University Press, 2021), 156. The textbook also discusses the dams built along the Missouri River and its tributaries, whose muddiness the Corps of Engineers failed to consider. Gavins Point Dam crossed the river by 1955; within sixteen years, it had trapped so much mud that the water backed up and flooded nearby towns, forcing Congress to set aside $18 million to buy out and relocate a flooded village. Absent some intervention, the reservoir will be filled within several centuries. Several smaller dams on the Missouri's tributaries will be filled within decades. See Hudson, *Flooding and Management*, 132, and Tyler J. Kelley, *Holding Back the River: The Struggle Against Nature on America's Waterways* (New York: Avid Reader Press, 2021), 141. The dams also mean less sediment flows into Louisiana, where, as we will see, there is an ongoing effort to use such mud to build more land.

225 **concrete pads:** "Concrete Mats," U.S. Army Corps of Engineers, https://www.mvn.usace.army.mil/Missions/Engineering/Channel-Improvement-and-Stabilization-Program/Revetment-Types/Concrete/.

226 **hardpoint:** David S. Biedenharn and Charles D. Little, Jr., *The Influence of Geology on the Morphologic Response of the Lower Mississippi River,* Mississippi River Geomorphology and Potamology Technical Program Report no. 17 (U.S. Army Corps of Engineers, 2018).

226 **"The river is not behaving":** Brien Winkley, *Man-Made Cutoffs on the Lower Mississippi River, Conception, Construction, and River Response* (Vicksburg, Miss.: U.S. Army Corps of Engineers, Vicksburg District, 1977), 202.

226 **the Mississippi "avulsed":** Roger T. Saucier, *Geomorphology and Quaternary Geologic History of the Lower Mississippi River* (Vicksburg: U.S. Army Engineer Waterways Experiment Station, 1994), 263.

227 **five hundred years ago:** Harold Fisk, *Geological Investigation of the Alluvial Valley of the Lower Mississippi River* (Vicksburg: Mississippi River Commission, 1944), 16.

227 **"a virtual laboratory":** Saucier, *Geomorphology,* 285.

229 **"the inexperienced traveler":** Quoted in Earth Search, *Cultural Resources Survey of EABPL Off-Site Borrow Areas* (U.S. Army Corps of Engineers, New Orleans District, 1994), 51.

229 **a half-million flood-prone acres:** Reuss, *Designing the Bayous,* 29.

230 **lift river stages as much as four feet:** Rodney A. Latimer and Charles W. Schweizer, *The Atchafalaya River Study* (Vicksburg: Mississippi River Commission, 1951), 4.

231 **nearly a million acres:** The size of the Atchafalaya Basin is surprisingly hard to pin down. See Reuss, *Designing the Bayous,* 5. This is the figure used by the Atchafalaya National Heritage Area.

231 **"just itching":** Quoted in McPhee, *Control of Nature,* 46.

231 **"There is":** Quoted in Reuss, *Designing the Bayous,* 214.

232 **"What we've done here":** Quoted in McPhee, *Control of Nature,* 21.

234 **Two lakes:** Reuss, *Designing the Bayous*, 154.

234 **"tiny fraction":** Saucier, *Geomorphology*, 285.

234 **"You know, Army Corps":** Quoted in Upholt, "River Is Under Control—For Now." Thanks to WWNO reporter Travis Lux for asking this question.

235 **already forty feet deeper:** Barnett, *Beyond Control*, 137.

235 **In March 1973:** Reuss, *Designing the Bayous*, 243.

235 **"Kid, it's gone":** Quoted in Barnett, *Beyond Control*, 146.

236 **"The problems in the river":** Winkley, *Man-Made Cutoffs*, 205.

237 **"We don't have":** Quoted in Joshua Alan Lewis, "Pathologies of Porosity: Looming Transitions Along the Mississippi River Ship Channel," *Urban Planning* 8, no. 3 (2023), https://doi.org/10.17645/up.v8i3.6954.

237 **a novel study:** Samuel E. Munoz et al., "Climatic Control of Mississippi River Flood Hazard Amplified by River Engineering," *Nature* 556 (2018): 95–98.

237 **nearly a quarter in Minnesota:** Friends of the Mississippi River and the Mississippi National River and Recreation Area, *State of the River Report 2016*, https://fmr.org/state-river-report.

237 **the Wabash River:** Thomas Höök et al., "Aquatic Ecosystems in a Shifting Indiana Climate: A Report from the Indiana Climate Change Impacts Assessment," *Aquatic Ecosystems Reports*, Paper no. 1 (2018): 4.

237 **the Minnesota River's:** Minnesota Pollution Control Association, *Our Minnesota River: Evaluating the Health of the River*, 2017, https://www.pca.state.mn.us/sites/default/files/wq-swm1-03.pdf.

238 **an average of 4.5 percent:** Eugene Turner, "Variability in the Discharge of the Mississippi River and Tributaries from 1817 to 2020," *PLoS ONE* 17, no. 12 (2022), https://doi.org/10.1371/journal.pone.0276513.

238 **The 2019 flood:** These details are drawn from National Weather Service, "Spring Flooding Summary 2019"; Jonathan Erdman, "Record Flooding in Nebraska, Iowa, South Dakota, Wisconsin, Minnesota and Illinois Follows Snowmelt, Bomb Cyclone," *Weather Service*, March 21, 2019; Ted Genoways, "River of No Return," *New Republic*, May 28, 2019; Laurent Belsie, "Nebraskans Talk Extreme Weather. Just Don't Call It Climate Change," *Christian Science Monitor*, April 6, 2019.

239 **Fourteen levee systems:** The figures are from John D. Holm, "2019 Missouri River Flood Event," presented at the Geotechnical Conference at the University of Kansas, 2019. Holm is the chief of the engineering department of the U.S. Army Corps of Engineers' Kansas City District.

239 **sent salinity plunging nearly to zero:** Boyce Upholt, "A Louisiana Spillway Helps Flood-Proof New Orleans. It's Killing Mississippi's Local Seafood Industry," *Counter*, September 16, 2019.

240 **on the order of $95 saved:** This figure appears in Mississippi River Commission, *Mississippi River and Tributaries Project: Without Flood Control, Nothing Else Matters* (2020), https://tinyurl.com/3rw9bs6e. The MR&T Project, according to this pamphlet, has prevented $1.5 trillion in flood damages. Such return on investment appears to be based on a false counterfactual: that everything now built behind the levees would have appeared there even if the levees had never gone up. (Social scientists have confirmed Charles Ellet's intuition that the "false security" of levees

tends to prompt increased investment—a clear case of moral hazard.) I asked a public information officer from the Army Corps of Engineers' Mississippi Valley Division to clarify the full cost of the MR&T Project, and got this reply by email: "$16.5B has been invested into the MR&T for planning, construction, O[perations]&M[aintenance], since 1928." When I asked if we could discuss how the costs were calculated—I was curious, for example, to know if it reflected local contributions—the spokesperson wanted to how I'd be using the information. After I explained that I was writing a book about the river, I never heard back. Other Corps of Engineers materials, published on the "Value to the Nation" portion of the agency website, indicate that the annual investment in the MR&T Project has averaged $560 million between 1950 and 2023. Investments were even higher in the project's first two decades. That would put the total investment well over $50 billion. The discrepancy is likely because in this case, each year's investment has been converted to 2023 dollars.

240 **"People's backyard pools":** Quoted in Danielle Dreilinger, "'Things Have Drastically Changed': Mississippi River Level Rebounds, Efforts Switch to Prevention," *USA Today,* March 12, 2023.

240 **three dredgeboats:** Morgan McFall-Johnsen and Ayelet Sheffey, "The US Army Has Been Dredging the Mississippi River 24/7 for 6 Months. The Drought Crisis That Grounded Barges and Unearthed Fossils May Finally Be Over," *Business Insider,* January 24, 2023.

240 **nearly four times the standard:** Andrew Stutzke, "Accurate Forecast," WQAD, November 15, 2022.

240 **as high as $20 billion:** "Mississippi River Crisis Losses Could Reach $20 Billion," AccuWeather, October 21, 2022.

240 **"This 'plume'":** River dredging, for the sake of ocean-going vessels, helped make such saltwater intrusion more likely; sea-level rise may contribute to future plumes. One bit of good news is that models suggest that the low-water conditions that marked these two years will become less low and less extended in the future. See Samuel E. Muñoz et al., "Mississippi River Low Flows: Contexts, Causes, and Future Projects," *Environmental Research: Climate* 2 (2023), https://iopscience.iop.org/article/10.1088/2752-5295/acd8e3.

241 **a thousand-year event:** K. B. J. Dunne et al., "Examining the Impact of Emissions Scenario on Lower Mississippi River Flood Hazard Projections," *Environmental Research Communications* 4, no. 9 (2022), https://iopscience.iop.org/article/10.1088/2515-7620/ac8d53. These labels can be confusing; they represent not a return interval but a statistical likelihood. A "thousand-year flood" has a one-in-one-thousand chance of occurring in a given year.

241 **"Results indicate":** J. W. Lewis et al., *Mississippi River and Tributaries Future Flood Conditions,* Mississippi River Geomorphology and Potamology Technical Program Report no. 28 (U.S. Army Corps of Engineers, 2019), ii.

241 **Indeed, weaknesses:** Lewis and Ashley, *Technical Assessment,* 11–12.

241 **The accumulation of mud:** The Morganza Spillway, too, has been impacted. The official trigger for its operation is a local discharge of 1.5 million cubic feet per second—but lately at that flow rate, the water is near the top of the control structure's gates. Since the gates cannot be opened if they are overtopped, in 2014 the Corps

altered its operational plan; the Morganza gates are now opened when the *forecasted* discharge reaches the trigger.

241 **And models show:** T. Mitchell Andrus and Samuel J. Bentley, "Capture Timescale of an Uncontrolled Mississippi–Atchafalaya Bifurcation with Future Lower River Strategy Implications," *Journal of Coastal Research*, forthcoming.

242 **"obliterated":** Dave Crane, an environmental resource specialist with the Omaha District, used this word at the 2022 meeting of the Upper Mississippi River Conservation Committee.

242 **$24 million per mile:** This figure, and the percentage of floodwaters stored, appears in Robert Jacobson, "Common Ground: Flood-risk Reduction and Conservation in Large-river Floodplains in Missouri," *Missouri Natural Areas Newsletter* 22, no. 1 (2022). Jacobson's essay implies that on the Missouri, where most of the historic floodplain is uninhabited but devoted to crops, it may make more sense not to build a series of setbacks but to remove some levees entirely, letting the river reach from bluff to bluff during floods.

12. Beautiful Country

Portions of this chapter have been published, in slightly different form, by *Hakai Magazine* ("The Controversial Plan to Unleash the Mississippi River," July 12, 2022) and *Bitter Southerner* ("Atchafalaya Mud," January 4, 2022). These stories go into slightly more detail about various aspects of these controversies.

245 **ecosystem restoration project:** See Restore the Mississippi River Delta, "Mid-Barataria Sediment Diversion: Our Best Shot to Turn the Tide on Coastal Land Loss," https://mississippiriverdelta.org/restoration-solutions/sediment-diversions/mid-barataria-sediment-diversion-this-is-our-best-shot/.

247 **the sole member of the council:** At the end of his term, in 2022, Blink did not run for re-election, citing a desire to focus on his ecotourism business. His replacement, Mitch Jurisitch, is an oyster harvester and an outspoken critic of the diversion.

248 **cranked up the river's:** R. Eugene Turner, "The Mineral Sediment Loading of the Modern Mississippi River Delta: What Is the Restoration Baseline?," *Coastal Conservation* 21 (2017): 868.

248 **a short piece:** E. L. Corthell, "The Delta of the Mississippi River," *National Geographic* 8, no. 12 (1897): 351–54.

249 **Today around half:** Rich Campanella, personal communication to the author, 2023.

249 **His first report:** Sherwood Gagliano et al., "Deterioration and Restoration of Coastal Wetlands," in *Proceedings of the 12th International Conference on Coastal Engineering,* (Washington, D.C., 1970), 1767–81.

250 **A subsequent study:** Aaron S. Bass and R. Eugene Turner, "Relationships Between Salt Marsh Loss and Dredged Canals in Three Louisiana Estuaries," *Journal of Coastal Research* 13, no. 3 (1997): 895. The removal of oil itself can cause subsidence, too.

250 **Powerful landowners:** *Jeffrey Meitrodt and Aaron Kuriloff, "Grounds to Sue," New Orleans Times-Picayune,* May 11, 2003.

251 **a near-unanimous no thanks:** Bob Marshall and Mark Schleifstein, "Losing Ground: Barely Making a Dent," *Times-Picayune,* March 6, 2007.

253 **"trembling land and swamp":** The quotations in this paragraph are from Craig E. Colten, "Cartographic Depictions of Louisiana Land Loss: A Tool for Sustainable Policies," *Sustainability* 10, no. 3 (2017): 764–65. The first quote is, in French, *"terre tremblant et marecageuses,"* and I have opted for a slightly different translation than Colten's version.

253 **"Malheur au blanc":** Quoted in Gwendolyn Milo Hall, *Africans in Colonial Louisiana: The Development of Afro-Creole Culture in the Eighteenth Century* (Baton Rouge: Louisiana University Press, 1992), 213.–

254 **runaway camps:** The Spanish word for a runaway like Juan San Malo was *cimarrón.* That translates as "wild," though with a particular emphasis: it is the wildness of an animal that was once domesticated but now has fled to the fields. Perhaps this word was meant to dehumanize these people, to reduce them to livestock, but if you appreciate the idea of wildness, it's easy to find something appropriate in the term. The runaways thrived in the margins, close enough to society to slip back and steal cattle, yet far enough away, in tangled-enough territory, to be difficult to track. The word is appropriate, too, because it expresses a truth about wildness. It's especially true of the wildness along the Mississippi River: it can be controlled for a while—domesticated—but eventually its unruliness will return. For the definition of cimarrón, see Erin F. Voisin, "Saint Malo Remembered," master's thesis, Louisiana State University (2008), iv.

254 **Historians estimate:** Benjamin Alexander-Bloch, "The Louisiana Seafood Industry's History, from Start in 1700s to Present, Discussed," *Times-Picayune,* February 28, 2013.

255 **system as carefully designed:** See Donald Davis, "Traînasse," *Annals of the Association of American Geographers* 66, no. 3 (1976): 349–59.

255 **more than half of Louisiana's muskrats:** Ari Kelman, *A River and Its City: The Nature of Landscape in New Orleans* (Los Angeles: University of California Press, 2006), 185.

256 **twelve feet of mud:** Benjamin Maygarden and Jill-Karen Yakubik, *Bayou Chene: The Life Story of an Atchafalaya Basin Community* (New Orleans: U.S. Army Corps of Engineers, 1999), 26.

256 **One Indigenous family:** Hannah Conway, "How Infrastructures Age: Engineering, Nature, and Environmental Justice in the Lower Mississippi Delta," Ph.D. diss., Harvard University (2023), 217.

256 **more than ten thousand miles:** R. Eugene Turner and Giovanna McClenachan, "Reversing Wetland Death from 35,000 Cuts: Opportunities to Restore Louisiana's Dredged Canals," *PLoS ONE* 13, no. 12 (2018): https://journals.plos.org/plosone/article?id=10.1371/journal.pone.0207717

258 **It's a net loss either way:** U.S. Army Corps of Engineers, *Final Environmental Impact Statement for the Proposed Mid-Barataria Sediment Diversion Project,* tables 4.2-3 and 4.2-4. Chris McLindon, a longtime oil and gas geologist, believes nearly all the land

loss in Louisiana can be explained by slips along underlying fault lines. (This is an idea that Sherwood Gagliano eventually embraced.) McLindon worries that a fault underlies the site of diversion, and he suggests that a "fault slip" could negate even the limited projected gains. In what he calls a "nightmare" scenario, it could prompt a crevasse, as has happened along other faults. In its final environmental impact statement, the Corps of Engineers acknowledged that faults are a contributor to land loss. Nonetheless, noting that scientists do not fully understand what triggers changes along fault lines, the Corps suggested it could not factor such issues into its analysis.

259 **They emerged some 7,500 years ago:** Paul Hudson, *Flooding and Management of Large Fluvial Lowlands: A Global Environmental Perspective* (New York: Cambridge University Press, 2021), 2.

260 **the river's next great renovation:** The state estimates that construction will take five years.

261 **the projected $3 billion price tag:** Around $300 million of that money will go toward mitigating damage. For the fishing industry, this will mean investment in new boat docks; in a marketing program for local shrimp, aiming to keep the local industry competitive against imports; and in oyster seed beds and the development of off-bottom oyster-farming industry. The state will elevate and build roads in communities that may see new floodwaters. Some money will go toward monitoring dolphin populations and intervening to assist sick dolphins.

261 **twenty feet of fresh silt:** Raphael G. Kazman et al., "If the Old River Control Structure Fails? The Physical and Economic Consequences," *Louisiana Water Resources Research Institute Bulletin* 12 (Baton Rouge: Louisiana State University, 1980): 12.

Epilogue: The Water of the Future

The Yazoo Pumps are a very complicated issue and difficult to summarize in a short space. For more details, see my "The Promise of the Yazoo Pumps," *Southerly* (2019), https://southerlymag.org/2019/10/10/the-promise-of-the-yazoo-pumps/.

265 **only a fifth:** Joshua Alan Lewis, "Pathologies of Porosity: Looming Transitions Along the Mississippi River Ship Channel," *Urban Planning* 8, no. 3 (2023), https://doi.org/10.17645/up.v8i3.6954.

265 **one of the largest rivers:** Andrew S. Lewis, "As the Mississippi River Swerves, Can We Let Nature Regain Control?," *Yale Environment 360*, September 5, 2023.

266 **a sand boil opened:** Tyler J. Kelley, *Holding Back the River: The Struggle Against Nature on America's Waterways* (New York: Avid Reader Press, 2021), 153.

266 **"the geologic evolution":** T. Mitchell Andrus and Samuel J. Bentley, "Capture Timescale of an Uncontrolled Mississippi–Atchafalaya Bifurcation with Future Lower River Strategy Implications," *Journal of Coastal Research* (2023), https://doi.org/10.2112/JCOASTRES-D-22-00127.1

268 **in a rare case of restraint:** Moira Therese Mcdonald, *"Political Economy of Agriculture in the Yazoo Delta: How Federal Policies Shape Environmental Quality, Livelihood Possibilities and Social Justice,"* Ph.D. diss., University of Minnesota (2010), 109.

268 **the last substantial component:** As proponents of the pumps often point out, three other "backwater" regions are included in the MR&T Project, where Mississippi floodwaters can push upstream into tributary mouths; all three feature pumps.

268 **"Ain't no money":** Quoted in Alexandra Watts, "Mississippi Delta Farm Workers Deal With Flooding's Aftermath," Mississippi Public Broadcasting, August 26, 2019.

268 **"The things I saw":** Hank Burdine, at a 2022 public meeting about the Yazoo Pumps in Rolling Fork, Mississippi. Burdine, a larger-than-life man of the river, passed away in 2023.

269 **a few feet higher:** In earlier iterations, the Yazoo Pumps were supposed to kick in once local waterways hit 87 feet. The new plan raised this threshold to 90 feet during the growing season, and 93 feet during the off-season. The pump's proposed capacity, meanwhile, increased from 14,000 cubic feet per second to 25,000.

269 **"human side":** Quoted in Mcdonald, *"Political Economy of Agriculture," 119.*

269 **"There's nothing new here":** Quoted in Courtney Ann Jackson, "New Plan Proposed For Yazoo Backwater Area and It Includes Pumps," *WLBT,* May 5, 2023.

269 **"hemispherically significant":** This phrase appears in a document titled "Yazoo Backwater: A Resilience Alternative" that was distributed by several environmental groups during the debates over the Yazoo Pumps.

270 **In a 2007 analysis:** U.S Army Corps of Engineers, *Yazoo Backwater Area Reformulation: Main Report* (2007), table 14.

273 **"a community-based action":** Jay Miller, *Ancestral Mounds: Vitality and Volatility of Native America* (Lincoln: University of Nebraska Press, 2015), 13.

Illustration Credits

viii–ix	Liz Camuti
xiv	Rory Doyle
14	Boyce Upholt
20	Liz Camuti
34	Boyce Upholt
37	Liz Camuti
48	Rory Doyle
55	Liz Camuti
70	Will Newton, Arkansas Department of Parks, Heritage, and Tourism
77	Liz Camuti
86	Boyce Upholt
108	William Alden
126	Rory Doyle
133	Liz Camuti
148	Boyce Upholt
162	Liz Camuti
164	Edward B. Madden, *Mississippi River and Tributaries Project: Problems Relating to Changes in Hydraulic Capacity of the Mississippi River* (U.S. Army Corps of Engineers, 1974).
172	Rory Doyle
183	Liz Camuti
196	Boyce Upholt
199	Liz Camuti
210	Liz Camuti

218 Rory Doyle

222 Harold N. Fisk, *Geological Investigation of the Alluvial Valley of the Lower Mississippi River* (U.S. Army Corps of Engineers, 1944).

228 Liz Camuti

233 Liz Camuti

244 Rory Doyle

252 Liz Camuti

264 Rory Doyle

Index

Page numbers in *italic* refer to photos and maps. Page numbers after 275 refer to notes.